Dragonflies & Damselflies *of the* Rocky Mountains

Dragonflies & Damselflies

of the

Rocky Mountains

By

Robert DuBois

Kollath+Stensaas
PUBLISHING

Kollath+Stensaas Publishing
394 Lake Avenue South, Suite 406
Duluth, MN 55802
Office: 218.727.1731
Orders: 800.678.7006
info@kollathstensaas.com
www.kollathstensaas.com

DRAGONFLIES &
DAMSELFLIES *of the*
ROCKY MOUNTAINS

Printed in South Korea by Doosan
10 9 8 7 6 5 4 3 2 1 First Edition

Editorial Director: Mark Sparky Stensaas
Graphic Designer: Rick Kollath
Illustrations by Rick Kollath

ISBN-13: 978-0-9792006-8-7

Table of Contents

Dedicated to

My wife Linda,

without whose support and encouragement
I could not have written this book

"A climb up the Rockies will develop a love for nature, strengthen one's appreciation of the beautiful world outdoors, and put one in touch with the Infinite."

—Enos A. Mills (1924)

Acknowledgements

My special thanks to my wife Linda, who once again was patiently willing to help me create time and space to research and write a book. I thank my parents, Robert P. and Malvina DuBois, for always supporting and encouraging my interest in the natural world, and my desire to study and write about it. And I thank my daughters Sarah, Danielle and Christina, for joining me on collecting trips, and for allowing me to share my enthusiasm for the natural world with them. Scott Wissinger accompanied me to several fine habitats in the vicinity of the Rocky Mountain Biological Laboratory (RMBL) at Gothic, Colorado, and shared many insights about high-elevation Odonata. Wendy Brown directed me to a number of great sites near RMBL. Scottie (Ruth) Willey also shared insights gained from many years of studying high-elevation Odonata in Gunnison County, Colorado. Jim Johnson and Dennis Paulson graciously supplied me with some much needed specimens for the illustrations. Since I am not much of a photographer, I am deeply indebted to those of much greater skill who provided examples of their fine work for this guide; a special thanks to all the photographers. A special word of appreciation goes to my frequent dragonflying companions Ken Tennessen, Bill Smith and Kurt Mead for their always-cheerful friendship. Dennis Pratt unselfishly helped ease an employment-related distraction that surfaced while this book was in its final stages—it's wonderful to have a friend you can count on in a pinch! Sparky and Rick, thanks again for your hard work and patience.

Bob DuBois
May 1, 2010

We wish to thank Bob for his full involvement in this project. His enthusiasm for damselflies and dragonflies has rubbed off on many folks… including us! Many hours afield spent wandering and wading were required for the creation of this book. Odonate photography is not an easy task, in fact, it is often a lesson in frustration. We give a hearty "well done," to photographers Ken Tennessen, Dennis Paulson, Ryan Brady, Mike Reese, Kurt Mead, Sid Dunkle, Karl Bardon, Giff Beaton and Jim Johnson. Enjoy the book.

The Publishers: Mark Sparky Stensaas, Rick Kollath
May 15, 2010

Parts of the Dragonfly

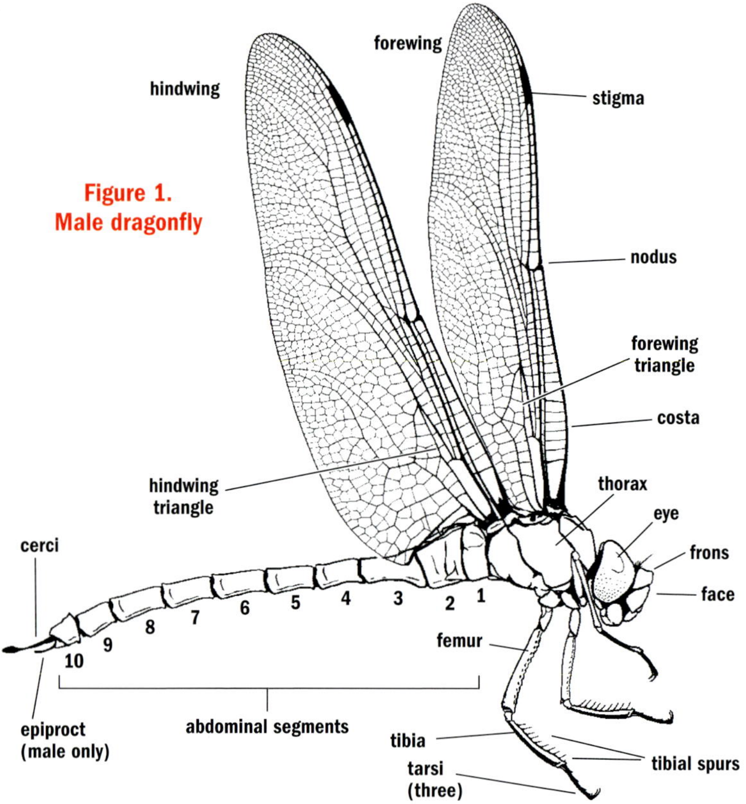

Figure 1. Male dragonfly

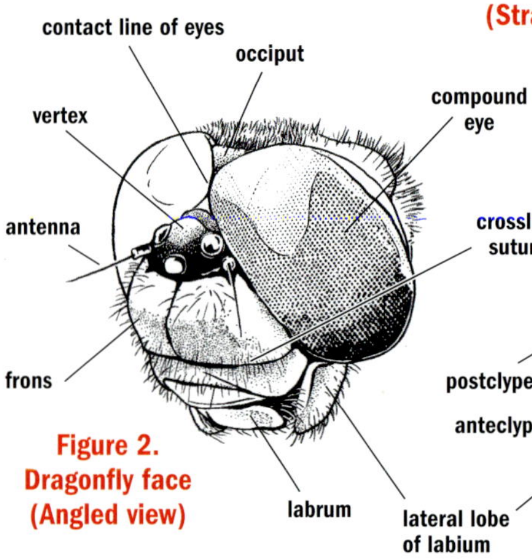

Figure 2. Dragonfly face (Angled view)

Figure 3. Dragonfly face (Straight on view with jaws open)

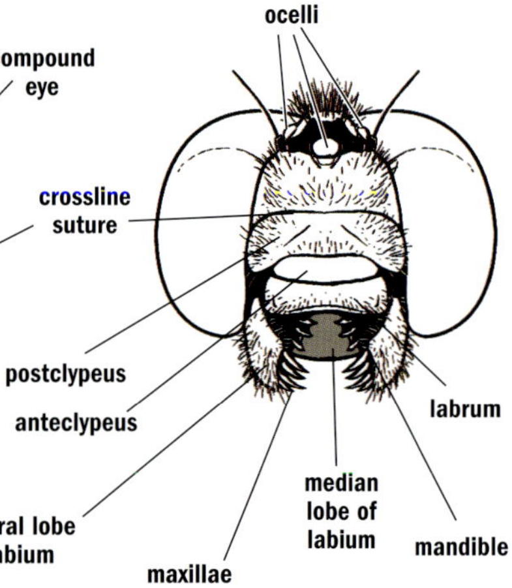

What are Dragonflies/Damselflies?

Remember when you were in high school and you learned about the biological classification system developed by Carl Linnaeus in 1758? Big groups were at the top of the chart that steadily got smaller toward the bottom: phylum and class, then order and family, then finally, the most specific categories of genus and species. Well, if you had a good science teacher, she or he probably explained to you why this terrifically successful system has remained in wide use for so long (with many modifications). One of the big reasons is that it gives us a framework for wrapping our minds around the staggering complexity and diversity of life. Otherwise, there would just be an enormous number of different critters out there, some more similar to each other than others, but we would have no frame of reference for organizing and discussing this diversity. So, how do dragonflies and damselflies fit into this system?

Dragonflies and damselflies are joint-legged arthropods (phylum Arthropoda), within the class of six-legged insects (class Insecta), within the order Odonata, which means "toothed" or "toothed ones" referring to their rapacious jaws. Other orders of insects with which you are familiar include the butterflies and moths (Lepidoptera), beetles (Coleoptera), bees, wasps and ants (Hymenoptera) and true flies (Diptera). Even though dragonflies and damselflies have the word "flies" at the ends of their names, they are not true flies. True flies should have the word "fly" as a separate word in their name, like house fly (*Musca domestica*).

Within the Odonata are two suborders of significance to us: the dragonflies (Anisoptera) and the damselflies (Zygoptera). The word dragonfly is used in many English-speaking countries as a coverall term to refer to both dragonflies and damselflies. I prefer to follow the distinctly North American tradition of using the term dragonfly only when referring to the Anisoptera and the term damselfly only when referring to the Zygoptera. Most naturalists use the term odonate when referring to both dragonflies and damselflies, and I will do so throughout this guide. People who study odonates are called odonatologists, or the shorter and for me preferable, odonatists. The common name "damselfly" comes from the French word *demoiselle* meaning "young mistress." I like to think that this refers to the slender, graceful appearance of damselflies. It is uncertain how the common name "dragonfly" came to be. Perhaps it arose from age-old but unfounded fears of these insects—or perhaps from their genuinely fierce facial expressions at close range. One theory refers to an ancient Romanian folk tale in which the devil turned a horse ridden by St. George into a giant flying

Parts of the Damselfly

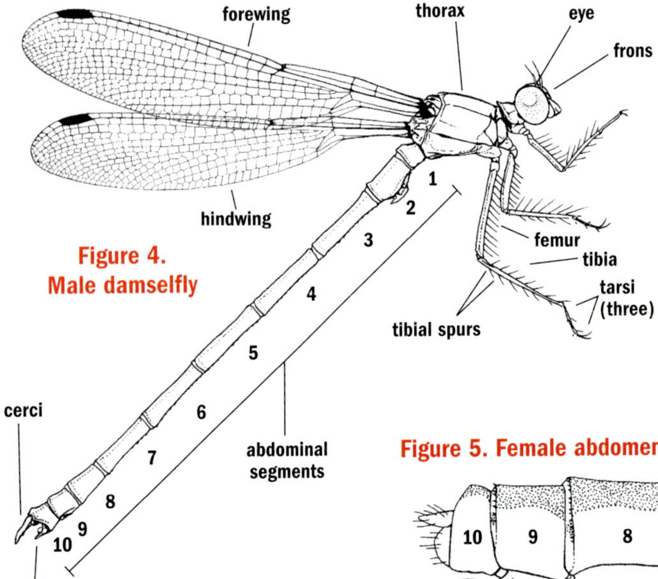

Figure 4.
Male damselfly

Figure 5. Female abdomen tip

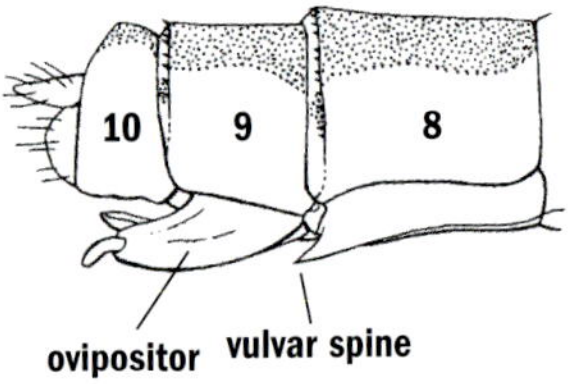

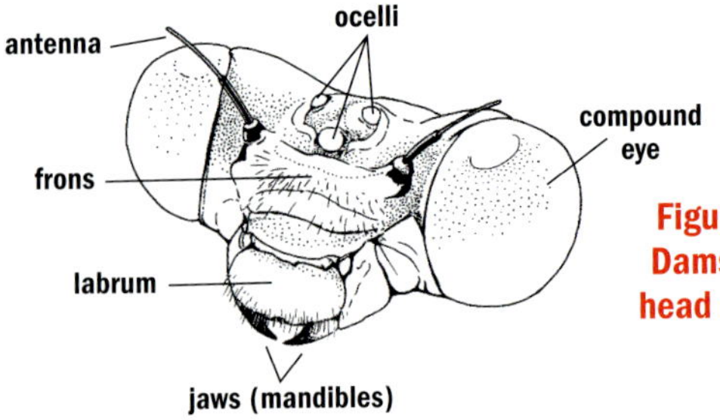

Figure 6.
Damselfly
head detail

Figure 7. Wing detail

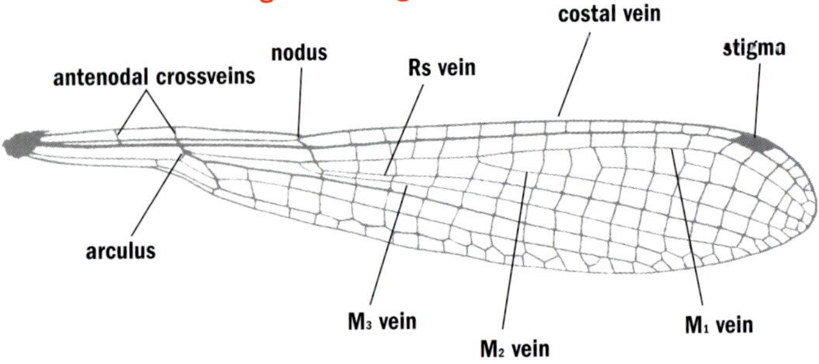

insect. Indeed, the Romanian word for devil is *drac*, but so is the Romanian word for dragon. So, I guess I'll have to leave this mystery unsolved!

Damselflies, like this Eastern Forktail, tend to hold their wings together up over their back.

To clarify how the two suborders differ, let's consider what their scientific names mean. Anisoptera means "unlike wings," referring to the hindwings of dragonflies being larger than the forewings and wider toward the wings' base. In damselflies, the forewings and hindwings are similar in size and shape. This similarity is alluded to in the name Zygoptera, which roughly translates to "paired wings," from the Greek term *zygon* meaning "yoke." Since they are in the same order, dragonflies and damselflies have a whole lot in common in terms of basic body plan and even many behaviors associated with reproduction, territoriality, emergence and feeding. Note the wing shape and other differences shown on the following chart to help you separate them.

Dragonflies

- Eyes in contact with each other *(separated by less than own width in clubtails)*
- Bulbous-headed *(head nearly as long as wide)*
- Stout build
- Strong, sustained flight
- Wings held nearly flat *(perpendicular to body)*
- Ovipositor non-functional *(except darners)*

Damselflies

- Eyes separated *(separated by at least own width)*
- Hammer-headed *(head much wider than long)*
- Long, slender build
- Weak, fluttery flight
- Wings held over back when perched *(or at roughly 45 degree angle)*
- Ovipositor present and functional

Diversity

There are 20 families of damselflies worldwide that contain about 2,700 named species. In North America north of Mexico we have five families with about 135 species. Here in the Rocky Mountains we have three families containing 40 species: two species of broad-winged damsels (Calopterygidae), six species of spreadwings (Lestidae) and 32 species of pond damsels (Coenagrionidae). There are eight families of dragonflies worldwide with about 2,900 named species. In North America north of Mexico we have seven families with about 320 species. Here in the Rocky Mountains we have six dragonfly families containing 59 species: 13 species of darners (Aeshnidae), five species of clubtails

(Gomphidae), one species of spiketail (Cordulegastridae), one species of cruiser (Macromiidae), eight species of emeralds (Corduliidae) and 31 species of skimmers (Libellulidae). Worldwide, many more species exist in both suborders that have not yet been named, especially in the tropics.

Dragonfly Parts

Figure 8. The widened end of the abdomen is called a "club." The Plains Clubtail has a substantial club.

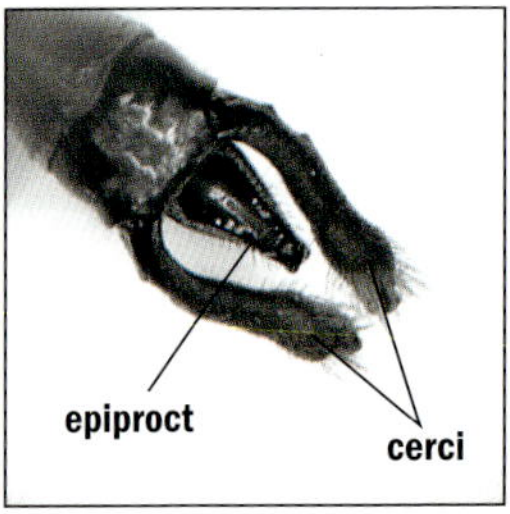

Figure 9. Male claspers at tip of abdomen are comprised of the lower epiproct and the upper cerci. They are used to hold the female's head during mating.

Figure 10. Female subgenital plate can be used for laying eggs. This emerald's is spout-like and perpendicular.

Odonates will be much easier to identify if you know a bit about their anatomy. I will use the terms "odonate" or "odonates" when describing parts that are similar for both dragonflies and damselflies. If a body part description applies only one suborder, I will refer specifically to that suborder. Fortunately, dragonflies and damselflies share many anatomical features, so you don't have to read two separate anatomy sections! Before we go any further, let's get our directions straight: **anterior** = toward the head (front) end; **posterior** = toward the rear end; **dorsal** = top side, toward the back; **ventral** = bottom side, toward the underside; **lateral** = toward one of the sides (what would be seen in side view); **basal** = toward the base of a structure (end toward the body); **apical** = toward the tip of a structure (end away from the body).

The primary divisions of the odonate's body are the **head**, **thorax** and **abdomen**, all covered with a hard (chitinous) **exoskeleton**. Being insects they have six jointed legs, two pairs of heavily veined wings as adults and ten abdominal segments. Let's take a closer look at these parts from front to back.

Head

The odonate head is large and has large compound eyes that contain 5,000 to 30,000 individual light-sensitive facets called **ommatidia**. The head is hollowed out at the rear and loosely connected to the thorax so that it can rotate freely when scanning side-to-side and up and down in search of prey. Therefore, odonates can see well all around them except perhaps for a small area to their rear. These features of the head and eye give odonates probably the best vision in the insect world. They can detect the movement of light at twice the speed we can. Damselflies have heads that are much wider than long, with moderately large eyes separated by a distance greater than the eye's width. This gives a damselfly a distinctly hammer-headed appearance. Damselflies have excellent depth perception and are well suited for slow flight among vegetation and accurate perception of small prey at short distances. Dragonflies have huge compound eyes, much larger than those of damselflies. Their eyes wrap around a substantial portion of the globular head and touch extensively along an eye seam in most of our families. However, the clubtail family has eyes that are completely separated like damselflies, but by a distance less than the width of an eye. The spiketail family has eyes that do not touch along an eye seam, but rather, barely touch at a point. Larger dragonflies have tens of thousands of ommatidia, which enables them to detect the slightest movement even at distances that may exceed 60 feet! This allows large, fast-flying species to be very effective aerial predators. Odonates can see color, and they can also see ultraviolet light and the plane of polarized light. Eye color can be quite striking, and useful in identification, with the upper part of the eye usually darker in coloration than the lower part. Three simple eyes called **ocelli** that appear to detect light intensity are also found on top of the head in odonates (see Figures 2, 3, 6). In dragonflies, the ocelli are located in an area just in front of the eyes called the vertex. Many species of damselflies have pale **eyespots**, called **postocular spots**, on top of the head that are useful in identification. These eyespots differ in size among species and may be connected by a pale, transverse line called the occipital bar. Behind the eyes lies a small, often triangular region called the **occiput**. Odonates appear to be deaf, so talk all you want in the field with your nature-loving friends!

The odonate "face" is made up of a number of plates separated by seams called **sutures**. From top to bottom the plates are known as the **frons**, **postclypeus** and **anteclypeus** (see Figures 2, 3, 6). The sutures may or may not be darkened and the face plates may have areas of color difference, all of which can be useful in identification. Beneath the anteclypeus is located the mouth, which is a complex but efficient "team" of parts that work together to make short work of their prey. The uppermost mouthpart (**labrum**) and the lowermost mouthpart (**labium**) are a bit like upper and lower lips. They hold the prey securely so the inner jaws can do what they do best—chew! The inner jaws are comprised of a pair of stout, wedge-shaped **mandibles**, that tackle most of the heavy duty chewing, and a pair of more slender **maxillae**. The **antennae** are small, bristle-like structures that may help the insect assess wind speed and direction, but are not used for smelling as they are in many other insect groups.

Thorax

The odonate thorax is a unique, backward-slanting box that is made up of three sections. The small front section, called the **prothorax**, controls movement of the head (functions as the "neck") and the front pair of legs. One of the signature features of odonates is that the two rear thorax parts (**mesothorax** and **metathorax**) are fused together to form the box-like **synthorax** (a.k.a **pterothorax**). This powerful "wheelhouse" contains the all of the flight musculature, to which the wings are directly attached, and it controls movement of the two hind pairs of legs as well. This flight musculature can comprise over a third of the entire weight of the insect! The structure of the thorax is skewed such that the legs are thrust forward and the wings backward, which aids in quick take-offs, the capture of aerial prey and for damselflies, the gleaning of prey from plant surfaces while in flight. However, this leg positioning makes walking difficult and awkward. The thorax is patterned in many of our species in ways that aid greatly in identification. A series of stripes form the foundation of these patterns in many genera. Our damselflies often have a relatively simple form of striping that consists, from top to bottom, of a middorsal dark stripe along the very top of the thorax, a pale shoulder stripe known as the **antehumeral stripe** just below that and a dark shoulder stripe called the **humeral stripe** below that (see Figure 11). The middorsal dark stripe may have thin pale line along the **middorsal carina**, which runs along the very top of the thorax. Some damselflies have no thoracic striping, others have the pale shoulder stripe reduced in the middle to a pair of dots, and still others

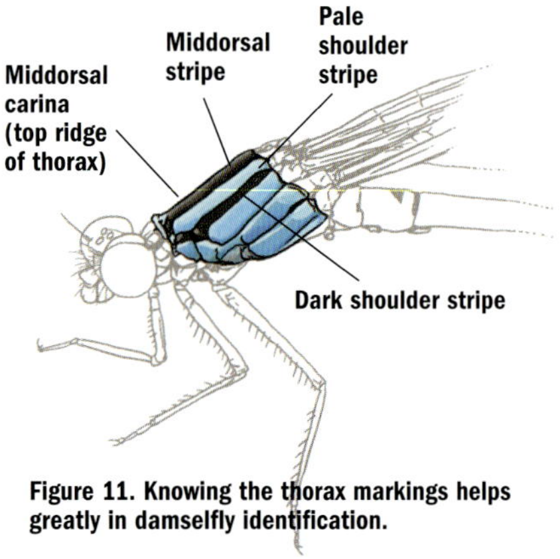

Figure 11. Knowing the thorax markings helps greatly in damselfly identification.

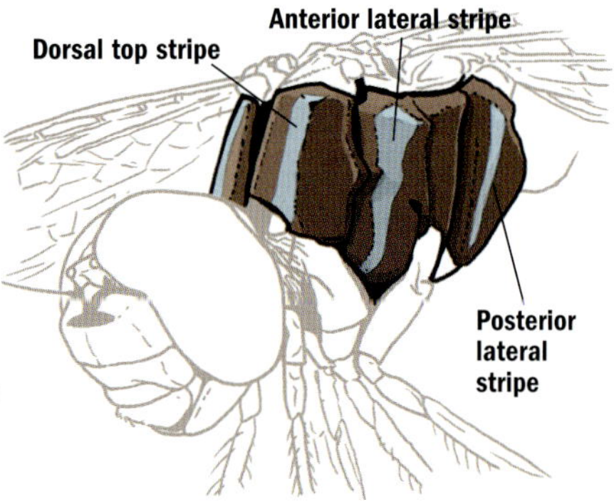

Figure 12. Thorax markings on dragonflies can be a great aid in identification—especially in the mosaic darners.

may have the thoracic pattern partially or wholly obscured by pruinescence (see section on coloration, page 21, or Glossary). Our dragonflies also have a thorax that may be unmarked, spotted, variously striped or obscured by pruinescence. Compared to damselflies, Rocky Mountain dragonflies with striping on the thorax have a more complicated pattern of stripes. It is beyond the scope of this guide to describe most of them in detail (but see Figure 12), but fieldmark arrows on the species pages will alert you to important stripes for identifying those species.

Wings

The wings of odonates are flexible structures with double membranes, laced with tubes within called veins that lend strength to the wing while at the same time allowing flexibility. Although they may look like pieces of cellophane or some other non-living material, wings are very much alive thanks to the veins. The veins contain nerves, some insect blood (**hemolymph**) and airways. Dragonflies and damselflies have the same major wing veins although the shapes of the wings differ between the suborders. Damselfly wings are generally narrowed (**stalked**) near the wing base whereas dragonfly wings are not stalked. In fact, the hind wings of dragonflies are widened near the wing base before narrowing abruptly to the thorax. All the wing veins of odonates—longitudinal veins, cross veins and even some areas of vein intersection—have names, and a thorough study of them takes time. In fact, a serious odonatist must learn at least three different wing venation naming systems because of disagreement among specialists about the evolutionary origin of the main veins! Fortunately for nature enthusiasts, learning just a few key veins and other wing parts will help immensely in identifying our Rocky Mountain species (see pg. 8).

Figure 13. Pay close attention to wing marks and patterns when out dragonflying. They often help in identification.

The strong vein along the leading edge of each wing is called the **costa**. It is distinctively colored in a number of dragonfly species, which will help you to identify them. A prominent feature of the wing is the

pterostigma, or **stigma** for short, which is a darkened, blood-filled cell of uncertain aerodynamic function near the tip of each wing along its leading edge. What is certain is that the shape and color of the stigma is often useful in identification. Near the middle of the wing along its leading edge you will see a slight notch called the **nodus** where several large veins join the costa. The nodus serves as a handy frame of reference in identification to locate wing markings or other veins. For example, the number of **antenodal crossveins**, which are located along the leading edge of the wing between the nodus and the wing base, are used to identify some groups of damselflies. The **triangles** (known as **quadrangles** in damselflies) of each wing are located about $^1/_5$th of length of the wing out from the base. The shapes and positions of the triangles relative to each other and to the **arculus** are important character states in separating several groups of odonates at the family level. The **radial planate** is a region of cells that is useful in identifying some Rocky Mountain whitefaces (genus *Leucorrhinia*) and meadowhawks (genus *Sympetrum*)—depending on whether it contains one or two rows of cells. A loop of veins called the **anal loop** is located near the base of the hind wing in most families of dragonflies. Its shape is used to separate emeralds from skimmers and to distinguish the group of clubtails known as snaketails (genus *Ophiogomphus*) from other clubtails.

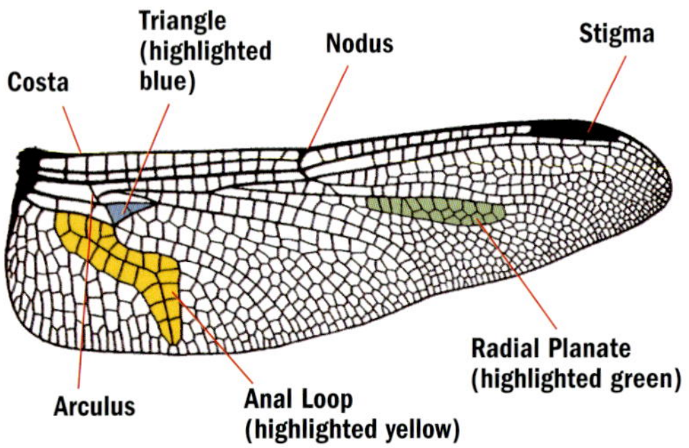

Figure 14. Knowing the main cells and other wing anatomy helps greatly in identification.

The wings of dragonflies and damselflies are directly attached to the muscles that move them, a condition called **paleopterous** (meaning ancient-winged). In contrast, most flying insects have the flight muscles directly attached only to the thorax and the wings move indirectly as a result of muscular contraction of the thorax. This latter condition is called **neopterous** (meaning modern-winged) and permits more rapid wing movement. This is one of the reasons why a house fly can flap its wings many more times per second than a dragonfly can. Odonates lack a hinge at the wing base and therefore cannot fold their wings flat over their backs. When perched, the wings of damselflies are usually held together vertically against, or slightly above, the body. However, the spreadwings (genus *Lestes*) rest with the wings partly spread at about a 45 degree angle to the body. The Taiga Bluet also perches with its wings slightly spread. Our damselflies never hold their wings out nearly horizontally and perpendicular to the body axis when perched as our dragonflies do (except for tenerals). Wings of odonates are usually clear, or hyaline, sometimes with a yellowish "wash" in some areas of the wing, but some species of skimmers, and our broad-winged damsels, have strikingly patterned markings on the wings that aid greatly in identification.

Legs

Like all insects, dragonflies and damselflies have six jointed legs (see Figures 1, 4). The long, upper leg section near the body is the **femur**. The next long section in the middle is the **tibia**. And the **tarsi** are comprised of three small sections making up the "foot," which is tipped with two small claws. Patterns of color on all leg sections can be useful in identification. Rows of spine-like **spurs** along the femur and tibia are useful in capturing prey. The relative lengths of these spurs are an

Telling Males from Females

Males

- "Bulge" visible on underside of abdominal segments 2 and 3 in side view; if reduced, sharp angle between segments 2 and 3
- Abdomen thin and slender
- "Club" prominent in species with abdominal club
- "Claspers" at abdomen tip have upper and lower structures *(not prominent in all groups)*
- No ovipositor or subgenital plate visible on underside of abdomen near tip
- Usually the more colorful gender when mature
- Hind wing may be angled (have corner) near base *(most dragonflies except skimmers and Common Green Darner)*

Females

- No "bulge" and no sharp angle between segments 2 and 3 in side view (smooth there)
- Abdomen stouter, bluntly tipped
- "Club" reduced in species with abdominal club
- "Paddles" at abdomen tip consist only of upper structures when prominent *(many groups no prominent structure)*
- Ovipositor or subgenital plate usually visible as bulge, spout or plate on underside of abdomen near tip
- Often drab; usually less colorful than mature male
- Hind wing never angled near base

important identification tool for pond damsels. The legs of nymphs are well developed for crawling and grasping the substrate. The legs of adult odonates are designed for perching and capturing prey in flight, but not for walking. Odonates do not need to use all six legs to perch—four legs, or even just two legs, are sufficient to maintain a secure grip. Thus, the two forward most legs are freed up for frequent preening of the head.

Abdomen

The odonate abdomen is slender, nearly cylindrical and studier than it looks. It always has ten **segments**, numbered 1 through 10, starting at the thorax (see Figures 1, 4). Patterns of color on these segments are often vitally important in identification, so it is well worth practicing how to count up from the base, or back from the tip (which I find easiest) to find the segment in question. Segment 1 is very short and may look like part of the thorax. Segments 2 and 3 are longer and modified underneath in males to hold the secondary genitalia, a complex array of reproductive organs that includes the **hamular processes**, **seminal vesicle** and the **penis**, among other things (see Figure 15). The secondary genitalia are usually easily visible on the underside of the thorax of the male and the presence of this "bulge" is the easiest way to determine gender—the female is smooth there. The shapes of the hamular processes are sometimes useful in identification of dragonflies, and when that is the case with Rocky Mountain species, these processes are clearly illustrated in this guide. In all Odonata, sperm is produced in testes near the tip of the abdomen at segment 9, but the male must transfer sperm to his own secondary genitalia prior to mating. This need, for a male to transfer sperm from one end of his abdomen to the other, is unique in the insect world. With his "tool kit" of reproductive parts the male is not only able to deliver his own sperm to the female, but he can also first physically remove or flush out any sperm of other males left from previous matings.

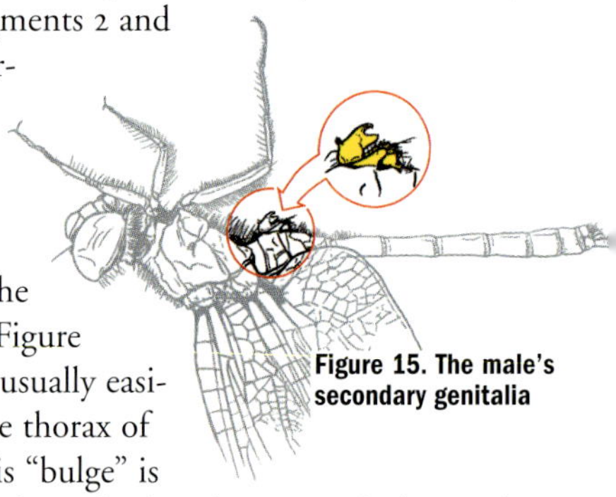

Figure 15. The male's secondary genitalia

To facilitate mating, male odonates have a set of **claspers**, also called **terminal appendages**, at the tip of the abdomen. The male uses these claspers to grasp the female before mating either by the prothorax (some damselflies and some dragonflies), or the mesostigmal plates of the synthorax (most damselflies), or the back of the head (most dragonflies). In dragonflies the claspers consist of an upper (superior) pair called **cerci** (singular **cercus**) and a single lower (inferior) structure called the **epiproct**. Female dragonflies have these parts reduced in size, except for female darners which have long, blade-like cerci that often break off while laying eggs. Male damselflies also have an upper pair of cerci, but the conspicuous lower part of their clasper arrangement is formed by the **paraprocts**. The shapes of these reproductive parts at the tip of the abdomen are crucially important in identification of many species. To avoid confusion I should add that damselflies have epiprocts, and dragonflies have paraprocts, but these structures are small remnants in those suborders. The cerci and paraprocts at the tip of the female damselfly's abdomen are also small, and serve no known function in reproduction.

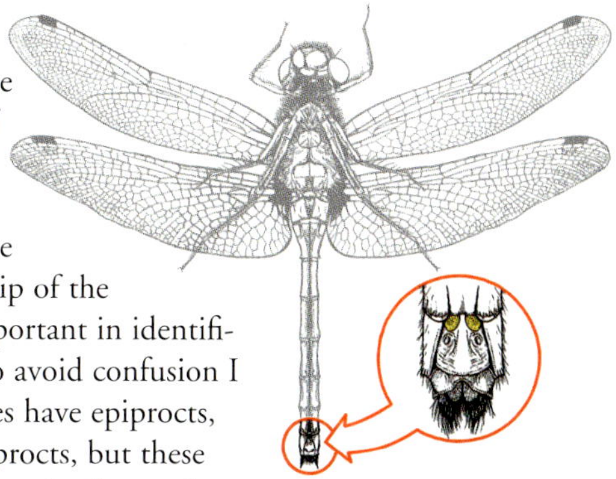

Figure 16. The shape of the female's subgenital plates is an important clue to identification of many species.

So remember:

***Male dragonfly claspers** = upper pair of cerci, and a lower single epiproct*
***Male damselfly claspers** = upper pair of cerci, and a lower pair of paraprocts*

The reproductive organs of odonate females consist of a genital aperture between segments 8 and 9, and on damselflies and darners, a well-developed and fully functional **ovipositor** (egg-laying structure) underneath segment 9. A functional ovipositor includes two or three sets of parts used to incise plant tissue and extrude one or several eggs into each cut. The shape and size of the ovipositor, and the presence or absence of a **vulvar spine** on the underside of segment 8, are useful aids in identification of females of many damselflies. Female dragonflies of most families have a **subgenital plate**, often referred to as a **vulvar lamina**, attached to segment 8 to protect the genital aperture and aid in copulation linkage (see Figure 16). The subgenital plate is peculiarly long in female spiketails and moderately long in some striped emeralds (genus *Somatochlora*) to facilitate egg-laying in wet sand or mud (figure 10). All of these reproductive parts are variable in size and shape among species and are therefore useful in identification. This discussion of the abdomen reminds me of the story of the young lad, who, asked by his teacher what the abdomen was used for, replied, "it contains the bowels, of which there are five—A, E, I, O and U!"

Odonate Biology 101

The life cycle of both dragonflies and damselflies includes a lengthy aquatic stage known as the **nymph** (also called **larva** or **naiad**), which is often a year or more in duration. Then, when weather conditions and season are just right, the nymph emerges from the water and undergoes a quick and absolutely stunning transformation into a winged adult that will only live for a month or two at most. Dragonflies are fiercely predatory both as adults and as nymphs, feeding for the most part on smaller terrestrial insects and smaller aquatic insects respectively. Despite their apparently gentle and serene demeanor, damselflies are just as ferociously predatory as dragonflies during both stages of life. In the long, slow nymphal stage, the goal is pretty much just to grow. During the brief, but frenetic, adult stage, the primary goal is to reproduce. But to reproduce effectively, young adults must feed, gain weight and mature sexually, before they are ready for the rigors of defending territories and competing for mates. Let's take a closer look at how this happens, first by peeking inside the bedroom door.

Mating

Dragonflies and damselflies mate only as mature adults. The required maturation period for newly emerged adults before they can mate lasts from a day to several weeks, and usually takes place away from the breeding site, often in nearby clearings and forest edges. When ready to mate, males usually congregate at **rendezvous sites** at or near the breeding areas where they space themselves to await the arrival of females. Males will usually try to mate with any female that enters the rendezvous area, and they can be a little sloppy about picking the right species. Fortunately, she gets to choose whether to mate or not, using subtle signals involving movements of her abdomen and wings. Some mosaic darners (genus *Aeshna*) and striped emeralds (genus *Somatochlora*) seem to gather at hilltops and ridgelines for mating, which means elevated areas can be great places to look for those groups.

Figure 17. A male River Jewelwing courts a nearby female with his wing display.

At the breeding site, males of some families actively defend territories while males of other families don't. In territorial species the idea is that the male perceives the territory he has chosen to be ideal for female egg-laying. Among Rocky Mountain groups, the broad-winged damsels, some dancers (genus *Argia*) and most clubtails and skimmers defend specific territories. Males of the darners, spiketails, cruisers, emeralds and a handful of clubtails and skimmers are not territorial, although they may chase other males when they encounter them. Males of most of our damselfly species are not territorial or especially aggressive toward other males, though when two male forktails (genus *Ischnura*) interact, one may chase the other from the rendezvous site. Males of some of our species wait well inland where they attempt to intercept females before they reach the rendezvous sites. Hagen's Bluet is known to employ this sly tactic. Males of other species will patrol for females along pond edges without being territorial or engaging in aggressive behavior with other males. As a rule, females arrive at rendezvous or breeding sites after the males and only when ready to mate.

Figure 18. A mating pair of Common Green Darners in the "wheel position." The male (blue abdomen) grasps the female behind her head with his claspers as he transfers his sperm packet to her.

To initiate mating, the male odonate uses his claspers to unceremoniously grab a female by the thorax or back of the head. If the female is receptive, the pair forms a "**tandem**" **linkage**, and they usually fly to a nearby perch if they are not already perched. Before mating, male odonates must translocate sperm from the testes on abdominal segment 9 to the seminal vesicle (part of the secondary genitalia) on abdominal segment 2. To do this the male arches his abdomen until the appropriate segments meet. Males of many dragonfly species perform the sperm transfer before engaging in tandem, but in some dragonfly and most damselfly species, the sperm transfer is done after tandem formation. Fully loaded, the male is now ready to initiate what is easily one of the most amazing spectacles in the insect world—formation of the **mating wheel**. If the female is ready to mate, she curls her abdomen forward so that the genital opening at the end of her abdomen engages the male's secondary genitalia. This produces a heart-shaped or circular configuration of their bodies commonly called being "**in wheel**" or "**in copula.**" While mating, the male supports both of them on a perch, the female's legs either being busy holding on to the male's abdomen (dragonflies), or simply not being used during this process (damselflies). Some species copulate in flight! How's that for multi-tasking? Copulation proceeds for several seconds (Common Whitetail) to several hours (some forktails), with several minutes being a good average. Much of this time is used by the male to remove any sperm of males that may have previously mated with that female (a fine example of sperm competition). During the last few seconds of copulation, he transfers his own sperm, depending on species either as free sperm or lumps of several hundred sperm (called **spermatodesms**), into her sac-like **bursa** (**spermatheca** in some species) where it is stored until she lays her eggs.

Mating between different species (hybridization) rarely occurs because of a variety of mechanisms that block either tandem formation or copulation. Nonetheless, males of many species can be shockingly undiscriminating, and may attempt to take in-tandem females of any species approaching the right size. Happily, female odonates are usually able to correctly identify males of their species by touch, and reject imposters.

Egg-laying and Guarding

The process of laying eggs is called **oviposition**. After mating, the male may leave immediately so that the female lays the eggs (oviposits) alone (called **unguarded oviposition**), or he may guard the female while she deposits the eggs. **Guarded oviposition** is typical for damselflies, some darners and skimmers, but most species in other dragonfly families do not guard. Two forms of guarding are used by the Rocky Mountain odonates that guard. Males of the broad-winged damsels release the female, but fly with her to the egg-laying site where he guards by hovering or perching nearby. This is called **non-contact guarding**. In this strategy, he chases away competing males that attempt to mate with the egg-laying female, and he remains free to mate with other nearby females himself. The most common strategy for guarding Rocky Mountain species is for the male to remain in tandem while the female deposits her eggs, called **contact guarding**. Contact guarding when egg-laying among some skimmers occurs in flight, thus differing from damselflies and darners that oviposit while perched. Either form of guarding enables the male to protect his genetic investment from being removed by competing males, and it allows the female to deposit eggs without being unduly harassed by other males. In addition, contact guarding serves to "encourage" the female to lay as many eggs as possible in the male's territory. The intensity and duration of male guarding seems to depend on the likelihood of interference from other males. Females of most species mate repeatedly during the several weeks of their adult life—some species do so every day. She does this even though she can store enough sperm from a single mating to fertilize several clutches of eggs without having to mate again. In a few species, notably some of our forktails (genus *Ischnura*), a single mating seems to be the rule, apparently because the females can store sperm for up to two weeks after insemination.

Figure 19. A female Black-tipped Darner inserts eggs into the stem of an aquatic plant in a quiet bog pool.

Most female odonates oviposit within 2 to 10 minutes after mating. Females oviposit in one of three ways: 1) into plant material (**endophytic oviposition**), 2) onto the surface of plant material (**epiphytic oviposition**), or 3) directly into the water or bottom substrate (**exophytic oviposition**). Females of all damselfly species use their ovipositors to deposit eggs endophytically, directly into the stems of rushes and sedges, into clumps of moss, into submerged plants or into pieces of floating plant material. The female uses her blade-like ovipositor to "saw" punctures into the plant tissue (either above or below the water level) and extrude one or several eggs into each puncture. Females of some damselfly species lay eggs into stems underwater; these may descend a foot or more below the surface and remain submerged for over one hour! Their supply of oxygen comes from a bubble or thin film of air that adheres to hairs on the body surface. Contact-guarding males may get dragged, kicking and screaming I imagine, partially or wholly under water as well, although they usually release their grip before their head gets pulled under. Females of most pond damsels rest horizontally on floating plant leaves when depositing eggs with the contact-guarding male standing rigidly upright as if "at attention," supported only by the grasp of his claspers on her thorax (see photo on page 89). I must confess that seeing males in this "**sentinel" position** often strikes me as amusing, although I doubt I could explain why. Occasionally, due to a predator attack, the abdomen of the male breaks off in this position, leaving the lower part still attached to the female and sticking straight up. Ouch! First time I ever saw this I thought I had witnessed some type of ghastly damselfly deformity.

Dragonfly females oviposit in a bewildering variety of ways that includes placing the eggs into plant material like damselflies (darners do this); onto plant material; into sand, muck or other bottom materials; directly into the water column; or onto the ground near a body of water where the next rain will flush the eggs into water. The shape of the female's ovipositor or subgenital plate offers a clue about the type of oviposition. Female darners with heavy-duty, blade-like ovipositors use them to cut into plant material, just like damselflies. Female spiketails have spike-like ovipositors modified for making holes in the bottom sand and silt of streams. Females with spout-like subgenital plates use them like ovipositors to plunge the eggs into wet sand or muck. Females with reduced subgenital plates lay eggs onto plant materials or directly into water because their subgenital plates don't function like ovipositors. Females of some dragonfly species oviposit while flying by tapping their abdomen tips here and there, releasing a few eggs or a cluster of them each time. Other species tap the abdomen and release eggs while perched. Released eggs sink to the bottom or may adhere to plant material. The numbers of eggs deposited by a female ranges from a hundred or so to many thousands depending on species and how long the female lives.

The Aquatic Nymph Stage

Dragonflies and damselflies spend the majority of their lives underwater as greenish or brownish, rather nondescript nymphs. Depending on their body shape and their other physical adaptations they crawl around, hide or burrow into the bottom material to hunt or ambush their prey. They live in nearly every imaginable type of freshwater habitat including rivers, streams, lakes, ponds, vernal (temporary) pools, marshes, swamps, bogs, fens and spring seeps. In the Rocky Mountains, the great majority of species live in still-water habitats, with few species found in rapid streams, especially at higher elevations. Nymphs in flowing waters have a generally flattened appearance (dorso-ventrally compressed) so they can squeeze into tight places and maneuver about without getting swept

Figure 20. Damselfly nymphs are slender and long-bodied compared to the bulky, oval nymphs of their cousins the dragonflies.

away by the current. This is a field guide to adult odonates, not nymphs, so I can't address nymphs in much detail here. But between you and me, the nymphs are every bit as interesting and just as much fun to learn about as the adults. Just tuck that thought away on your back burner for now.

The nymph stage begins, quite naturally, with the hatching of the egg. Odonate eggs are round, oval or elongate, and are covered with a sticky "jelly" in most families. The eggs hatch rather quickly, usually from 12 days to five weeks after being deposited, depending on air and water temperature, season and species. The eggs of most of our spreadwings and some of our meadowhawks (genus *Sympetrum*) overwinter—all other Rocky Mountain odonates overwinter as nymphs. When an egg hatches, a tiny pronymph is released. Within just seconds or minutes, the exoskeleton of the **pronymph** is rent apart and the first true nymphal stage bursts forth. Because the exoskeleton of the nymph has little elasticity, the only way for the nymph to grow is to shed the old skin and replace it with a slightly larger one. As the nymphs feed and grow, Rocky Mountain species progress through about a dozen **molts** on average (range from about eight to 16 molts depending on species) before they are ready to transform into adults. Each period between molts is called an **instar** (the term **stadium** is almost, but not quite synonymous). For older nymphs, it is fairly easy to determine which instar the nymph is in by observing the relative size of the wing pads, which enlarge and lengthen incrementally from instar to instar. If the hind **wing pad** (**sheath**) is about as long as, or slightly longer than the nymph's head is wide, then the nymph is in its **final, or terminal, instar**. The nymph stage typically lasts a year or two, but sometimes up to four or five years, or even longer. A general rule is that larger species, and species living in more northerly environs, take longer to develop. Overwintering nymphs can feed, but probably don't grow much then or molt. The Taiga Bluet nymph is known to overwinter embedded in ice! Now there's a nifty way to avoid getting preyed upon! Some species of our spreadwings are adapted to live in temporary ponds; in these species the nymph stage may be completed in five months or less, before the ponds dry up.

Tireless hunters, odonate nymphs lurk on stems of standing aquatic plants or in beds of submersed vegetation, actively pursuing smaller insect prey or even tiny fish. To grab their prey, they are equipped with an elongate, hinged lower jaw (labium), which they are able to extend out with lightning speed (about 1/100 second), latching onto the prey with their raptorial **labial hooks** (Figure 21). Like so many things peculiar to the Odonata, this highly modified mouthpart is unique in the insect world. Although capable of striking fear into the hearts of a vast array of small aquatic critters, odonate nymphs do not bite or harm humans in any way.

While obtaining dinner is important to an odonate nymph's survival, avoiding being dinner for some scary creature larger than itself is clearly a priority as well. Many animals dine on odonate nymphs, including water birds, sunfishes, black basses and even larger odonate nymphs. Ingenious studies by Mark McPeek of Dartmouth College and colleagues have shown that ponds with sunfishes and black basses (family Centrarchidae) tend to have different damselfly species than fishless ponds. It turns out that some species of damselfly nymphs try to escape from predators by swimming away, while others are content to crawl away and hide. The strategy of swimming away works well when escaping from insect predators, but not so well when the predators are fish. In a somewhat similar vein, many species of dragonfly nymphs have stout hooks on the dorsal surface of some of their abdominal segments and prickly lateral spines on some segments as well. Research has shown that among species that can live in habitats either with fish or without fish, those populations coexisting with fish tend to have longer lateral spines and stouter dorsal hooks than those populations living in the absence of fish. Evidently, the pricklier the nymph, the less likely a fish is to consummate the meal. This "use it or lose it" adaptive strategy is known as a form of induced morphological defense. Pretty cool stuff, eh?

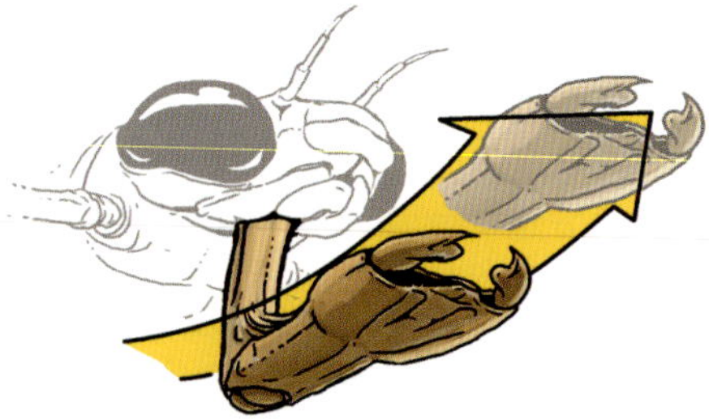

Figure 21. The aquatic nymphs of dragonflies are ferocious predators. Extendable hinged lower jaws (labium) can strike and grab prey before they know what hit them.

Damselfly nymphs differ from dragonfly nymphs in several ways. Damselfly nymphs are more slender, and unlike dragonfly nymphs, their head is usually wider than the thorax or abdomen. They also have three, slender-leaf-shape "gills" called caudal lamellae at the tip of the abdomen that help them breathe under water and swim gracefully in side-to-side fashion. Dragonfly nymphs lack **caudal lamellae**, have a "pyramid" of five, stout-pointed appendages at the tip of the abdomen and, wording this as politely as I can, breathe through gills inside the rectum. They can also powerfully "shoot" water out of the anus in jet propulsion fashion to move quickly from one place to another without any leg movement.

Thus the dragonfly enters upon a more nobler life
than that it has hitherto led in the water,
for in the latter it was obliged to live in misery,
creeping and swimming slowly,
but now it wings the air.
—Jan Swammerdam (1637–80)

Emergence

After months or years of clambering about underwater, the nymph is freed from the shackles of this ignoble existence in one grand moment of emancipation. Almost instantly it becomes one of the most graceful, elegant and masterful flying creatures under the sun. In the days before this stunning transformation the nymph goes through many changes; it becomes more sluggish, stops feeding, spends more time near or at the surface of the water and the wing pads darken and begin to swell. The mouthparts and other nymph-specific structures become dysfunctional because the new adult (**pharate adult**) is now forming while imprisoned inside the nymph's exoskeleton. The large eyes of the adult can often be seen through the nymph's cuticle on top of the head, and some wing veins can be seen inside the wing pads. Just before **emergence** (technically the last **ecdysis**), the nymph rests at the water's surface with head and part of the thorax exposed and begins breathing through **spiracles** that have opened up on the sides of the thorax.

The nymph must be fully mature (in its final instar) before it can undergo the form of incomplete metamorphosis we commonly call emergence. **Incomplete metamorphosis** simply means that odonates, unlike some other insect orders, lack a pupal stage. Many species emerge during the night or wee morning hours, but depending on species, weather conditions and just the variation within species, emergence can take place at just about any hour. Many clubtails and most damselflies emerge late in the morning or during midday. At higher latitudes and elevations, even night-emerging species may emerge during the day because the nights are so cold. Most of our damselfly species climb from just a few inches to a foot or two above the waters' surface on aquatic plant stems and emerge in a vertical position. Some species of dancers climb onto logs or rocks and may emerge in a horizontal position. Dragonflies emerge in similar fashion but some species may crawl a bit further from the waters' edge and depending on species, can emerge in vertical, oblique or horizontal positions. Some species have been known to crawl over 100 feet overland to an emergence perch, but the vast majority of emergences occur within just a few feet of water.

Emergence perches can be any structures near the waters' edge—plant stems, rocks, exposed banks, tree trunks, boat docks, bridge abutments and, well, you get the picture. In semiarid areas, some individuals fail to find an emergence perch and keep walking until they die.

To begin emergence, the nymph gains a secure grip on the emergence perch with its **tarsal claws** and rests for a few minutes (longer in larger species) while drying off. Then the thorax begins to bulge and splits longitudinally along its top. The top of the head also splits from side to side. These splits meet and then the thorax of the new adult, followed by its head, push out from this opening in the nymphal exoskeleton (**exuvia**, plural **exuviae**). Now free, the head and the thorax hang backwards for a while, usually about twenty minutes, while the new exoskeleton and the legs begin to harden. When the legs are strong enough, the new adult suddenly bends forward and grasps the emergence perch or its own exuviae and pulls out its abdomen. The abdomen extends with air pressure from the gut and the tightly folded wings unfurl and expand as blood (hemolymph) is pumped into them. Full expansion of the wings and abdomen takes up to an hour before flight can be attempted.

The newly emerged odonate, now called a **teneral**, is soft-bodied and pale yellow or light green in color with no dark markings. Tenerals gain color quickly, but most species don't acquire full adult coloration until

Figure 22. Emergence is a process....Metamorphosis of the aquatic nymph to the terrestrial adult dragonfly takes place on land. After clamping on to a log, grass stem, dock, etc. the nymphal case splits open and the thorax and head emerge.

at least a few days after emergence. Initially the flight is feeble and the wings glisten and shimmer in the sun. Right after emergence the teneral is very vulnerable to any predators along the shore—small mammals, birds, frogs, spiders, predatory insects and even the carnivorous plants (like sundews) it may land on. Many spring species synchronously emergence in huge numbers so that some of them make it through the gauntlet of predators. First flights are away from water as the tenerals disperse from the breeding site to feed, strengthen, mature sexually and gain weight in nearby clearings, meadows, roadsides and forest edges. But gaining weight has its limits—small dragonflies and damselflies are not babies, they are simply fully grown adults of small species. Emergence of most individuals of a species usually occurs over a two- to four-week period, although some forktails extend emergence over several months.

The exuvia remains on the emergence perch, often for a week or two, until it is dislodged by wind, rain or high water. Exuviae retain the form of the mature nymphs very well, and for that reason, can usually be identified to species using dichotomous keys. Collecting exuviae along shorelines is tons of fun and is perhaps the best way to document successful breeding of a species at a site (adults could have flown from elsewhere). The downside is that special equipment is needed to identify nymphs and exuviae (stereomicroscope, light source, expensive keys,

Figure 22 continued. After a wriggling struggle, the abdomen is finally pulled free, wings still crumpled over its back. Blood (hemolymph) is pumped into the wing's veins until completely unfurled. This may take about an hour. The newly emerged odonate is a teneral.

etc.), and the process of learning how to do it is, well, challenging. But don't let me talk you out of it! Another thoroughly enjoyable pastime is rearing nymphs. If you collect some mature nymphs (with big wing pads) and put them in an aquarium with some water and emergence perches for them to crawl up on (sticks or pieces of screening work well), you will be able to watch the emergence process. You won't even have to feed them unless they are young. Oh, and don't forget to put some netting over the top of the aquarium or you might have adults flying up and down your hallways!

Coloration

One of the reasons why dragonflies and damselflies are so enjoyable to watch is that they come in such a striking variety of resplendent colors—just about every color in the rainbow! Our Rocky Mountain species display various hues of yellows, orange, crimson, sapphire blues, shimmering greens, dark purples and rich magentas, black and a wide spectrum of browns and grays. Some species have areas of metallic green or bronze. Colors and patterns play a huge role in identifying all our species. Patterns of coloration that contrast with the surrounding areas come in the form of stripes (narrow and elongate, running with the body), spots (roundish and usually small) and transverse bands (often called rings or annuli if they encircle a structure).

Learning to recognize our 99 species of odonates by colors and patterns is a bit more challenging than would initially appear. This is because males and females often do not look alike (sexual dichromatism). Also, tenerals of all species are much paler than mature adults, and lack dark markings. Males of most groups are usually more brightly colored than their females. Identifying females to species can be quite tricky, especially with pond damsels, as there can be male-like forms and species that become grayish or bluish as they age (**pruinescence**). Pruinosity forms from a waxy substance that exudes from the cuticle. It is not limited to females; in spreadwings pruinosity is most pronounced in males. Male Powdered Dancers become highly pruinose to the point of becoming nearly white as they age. In some skimmers, both genders are pruinose. Further, the bright blues of some pond damsels and darners can change to dull purple in response to temperature decreases or darkness.

Parasitism

As you begin to observe odonates closely you will soon notice that many have small orange or red egg-like bodies attached to the underside of their thorax or abdomen. These are not the eggs of a female, but rather are water mite parasites (most commonly genus *Arrenurus*) that settle on the nymph while underwater. When the nymph emerges, the larval mite is able to quickly crawl off the exuviae and transfer his boarding pass to the teneral adult. The mite now begins the parasitic stage, attaching to the new adult odonate and sucking its body fluids for at least three or four weeks. It will then drop off back into the water to complete its life cycle. Parasites do not usually kill their hosts, but a heavy parasite load may hamper the feeding and reproductive success of odonates. Pond damsels and smaller skimmers are usually the most heavily parasitized odonates. A variety of other invertebrates can parasitize odonates including parasitic wasps, some of which can fly underwater to find the eggs (I couldn't make this stuff up if I tried), and a species of biting midge (family Ceratopogonidae) that likes to suck blood from a wing vein even when the dragonfly is in flight!

Figure 23. Note the orange-red hitchhikers on the underside of this damsel's thorax. These are parasitic mites.

Feeding and Flight

Adult odonates feed in two ways, the most common being the capture of flying insect prey on the wing using their forward-facing and spur-adorned legs kind of like a net. Some damselfly species also feed by gleaning small insect prey off of the surfaces of foliage. Pond damsels are especially good at gleaning. Odonates feed opportunistically on whatever prey species are available, showing little selectivity for particular prey. A wide variety of flies are often the staple, with side dishes of caddisflies, small moths and other insects. Odonates will certainly eat mosquitoes, both during the odonate nymphal stage and as adults, but it is doubtful that they take enough to noticeably reduce populations. Once secured, the prey is quickly consumed in big chunks either in flight or after it is carried to a nearby perch. Contrary to the beliefs of some, odonates do not sting nor do they bite hard enough when handled to cause any serious pain to humans. That is, unless a really big one gets you in a sensitive area when removing it from your net, but even then, the bite is harmless. They won't come after you to bite you!

It is the remarkable and complex flying ability of odonates that allows them to be very effective aerial predators. Dragonflies are master aerialists, blending their capacity for blazing speed with amazing maneuverability. Larger dragonflies can attain burst speeds that may reach or possibly exceed 30 miles per hour. Amazingly, they can carry off prey that weighs 15 times more than they do! Damselflies excel at being quick, precise and extraordinarily maneuverable. Odonates can fly in virtually any direction, even straight up, down or backwards. The strong, flexible odonate wing generates a tiny vortex of turbulence about the tip. Odonates evidently use the turbulence they create to maneuver. Their forewings and hind wings usually beat about 180 degrees out of phase, but they can also move all four wings synchronously and they can control each of the four wings independently. Odonates possess several flight advantages over other insects including an unusually high ratio of flight muscle to body mass and very low **wing loading** (the ratio of body mass to wing area). Dragonflies generally forage above the vegetation where their vision and speed allows

Figure 24. Orbweaver spiders and their large, sticky webs are a definite hazard to low-flying damselflies.

them to spot and overtake smaller insects. Damselflies are best suited to foraging low and slowly among vegetation, where their excellent vision allows them to home in on small prey at short distances. The habit of most damselflies to fly low, both amongst vegetation and over water, probably serves to reduce the risk of their being eaten by birds.

Migration

The kind of migration that insects engage in consists of one-way movements from one place of residence (such as a breeding site) to another. Insects don't live long enough to move from one place of residence to another and back again, as do some birds, marine turtles and some fishes. Migration in odonates is imperfectly known and the movement patterns of some species are poorly understood. Some species seem to move just enough to find areas of less intense competition for resources and mates. A few species in the darner and skimmer families regularly engage in long-distance migrations. One champion of this behavior plan is the Common Green Darner. Spring migrants of this species fly north to reproduce in northern areas. Their progeny (the nymphs) develop with explosive rapidity, often in one summer, and head south in swarms shortly after they emerge in late summer. The aptly named Wandering Glider is another committed wanderer, known for crossing ocean expanses of thousands of miles. In North America they seem to do the same north/south thing as the Common Green Darner. A third Rocky Mountain wanderer, the Variegated Meadowhawk, exhibits the unusual pattern of dispersing from west to east in the fall and from south to north in the spring. Other Rocky Mountain species and genera with "itchy wings" include the Familiar Bluet, many of the mosaic darners (genus *Aeshna*), the Twelve-spotted Skimmer, both saddlebag species (genus *Tramea*) and the Spot-winged Glider. The Familiar Bluet disperses enough to be very effective at finding and colonizing newly created or temporary habitats. Adult dragonflies will often travel several miles from the breeding site to feed and mature, but large-scale migrations are limited to a relatively small number of species.

Odonate Observation

Why watch dragonflies and damselflies? Well, there are a bunch of good reasons, but among the best are that they are unique and often beautiful little creatures, fascinating in their behaviors and unlike so many other animals, they are often easy to approach. Most species will offer the patient and careful observer a thorough look at all their sundry activities. Happily, most odonates are active right when many of us like to be out and about—late mornings and afternoons on sunny, warm days. I like the way Kurt Mead put it in *Dragonflies of the North Woods*, "the best time for dragonfly observation usually occurs after a long, late breakfast, complete with an extra cup of coffee." Indeed, they are card-carrying sun-worshipers—to the point that even a few dark clouds passing over will cause most of them to disappear so fast you'll wonder what happened. They don't actually disappear, but most will become inactive and perch in trees, bushes or other vegetation in the absence of direct sunlight. However, a few of our species are **crepuscular**, meaning they fly into dusk. At the other extreme, too much heat will also cause them to become inactive. At these times they will seek shade, or some species will point their abdomen towards the sun (called **obelisking**) to minimize the solar radiation hitting their bodies.

All ages enjoy the hobby of "dragonflying." Learning to handle live insects is just one fascinating aspect.

For the beginner, a good way to get started in dragonflying is simply to pull on a pair of boots (or wet wade if you like), grab this field guide and head out to a nearby pond, lake, river or wetland on a sunny afternoon. My initial advice is to keep it simple. Allow yourself to enjoy the pleasures of discovery and exploration! Don't be afraid to hop right into their habitats. It is much more satisfying to immerse yourself in the dragonfly's world than just watching

Obelisking is an easily observed dragonfly behavior.

their behavior from a distance. Perhaps the best tip I can give you is to move slowly and pause often. You'll see much more that way. Observe the colors, sizes, behaviors and habitats of the odonates you see flitting about you. Try to guess what they're doing and why. See if you can tell what family or genus they belong to. If you enjoy photography, you may want to learn how to photograph your colorful subjects.

Habitats and Seasons

The U.S. Rocky Mountains is a vast region that encompasses the interior cordillera of the western United States—the Southern Rockies alone cover roughly 50,000 square miles. This vast landscape is home to a very rich and wonderful tapestry of aquatic habitats in which odonates can thrive. However, it is hard to find words to adequately and accurately describe the habitat diversity and flight seasons for odonate watching across such a vast and variable landscape. The hydrology of the Rockies is dominated by snow accumulation in winter and snowmelt run-off during late spring and summer. Generally speaking, headwater streams throughout the Rockies have elevated flows in spring and early summer as the snows melt off the mountains. These highly variable flows, with their coarse and inhospitable bottom substrates, coupled with short growing seasons, make life hard on nymphs. So streams at higher elevations will usually have few if any odonates. Further downstream, streams and rivers widen in the montane valleys and river-loving species become more common. The U.S. Rockies contain tens of thousands of alpine, subalpine and montane lakes that form the "bread and butter" habitats for Rocky Mountain odonate watchers. These lakes are highly variable depending on latitude, elevation and geology, but what they have in common is they all produce odonates,

Quiet bog pools are the preferred breeding grounds for many dragonfly species in the Rockies. Courtship, mating and egg laying will take place here. Their nymphs will develop on the pond's bottom, eating voraciously until it is time to emerge.

except for the highest altitude alpine lakes. Wetlands in the form of marshes, swamps, bogs, fens, spring seeps and warm springs are liberally scattered throughout the Rockies, and they too produce plenty of odonates. These wetlands are most commonly associated with seepage slopes, river floodplains, kettle depressions and other landforms caused by glacial retreat.

Good dragonflying habitats

Good dragonfly and damselfly habitats; from top to bottom: Beaver ponds, streams at lower elevations (up to about 8,000 feet elevation in Colorado), fens and wet sedge meadows (this is a rare iron fen.)

Odonates can be found just about anywhere throughout this highly varied landscape, particularly right after emergence when they scatter from breeding sites. But your best chance of finding lots of action is to go to favored breeding areas. Well-vegetated pond and lake shorelines will reliably harbor a good number of species. Marshes and swamps will also hold their share. Rivers and streams in the Rockies usually have different and fewer species than do still-water habitats. But don't miss out on flowing-water habitats because the species there are strikingly colored and absorbing to watch. Be especially alert for Pale Snaketails, River Jewelwings and a variety of dancers along rivers and streams. As mentioned earlier, high-elevation streams (above about 8,000 feet in Colorado) present too harsh an environment for odonates, so confine your river odonate-watching trips to foothill, montane and lower subalpine areas. Bogs and fens will often have darners, emeralds, spreadwings and pond damsels as will the many small kettle ponds that dot the landscape in many areas of the Rockies. Like birds, some odonate species are adapted to very specific habitats, like warm springs or acidic fens, whereas other species are more cosmopolitan and are likely to found at almost any body of water. So if you are looking for a particular species, go to the best example of that species habitat you can find, based on its preferred habitat as described in this guide. Concentrations of dragonflies are often seen chasing after prey (known as **hawking**) over fields or roads. These hawking groups are often comprised of only a few closely related species, and they may be challenging to net.

A number of species will be active each month from June through September in the Rocky Mountains. However, the species present in the spring will be different than those present in the summer, or autumn, especially in the Foothills and the Montane Zone. Of course there will be some overlap because some species have longer flight periods than others. At the higher elevations of the lower Alpine and Subalpine zones, the flight season is shortened to the point that considerable overlap occurs between early and late season species. And, unseasonably warm or cold weather can either lengthen or shorten the flight season at any elevation. These sources of variability make simple statements about flight periods to be unhelpful. Throughout the Rockies, the diversity of species generally decreases as latitude, and especially elevation, increase. I say that to inform, but do so somewhat reluctantly, because it is the challenge of finding the uncommon, and the beauty and interesting behaviors of individual species, not just the sheer number of species,

Poor dragonflying habitats

Inhospitable habitats where you would be hard-pressed to find any odonates; from top to bottom: High Alpine shallow ponds, Great Salt Lake, Alpine mountain rapid-flowing streams.

that makes dragonflying so enjoyable. Some of my most satisfying days in the field have been in areas where diversity was not particularly high.

Gear and Equipment

As your interest in odonates blossoms, you will find a few tools to be helpful. At the very top of this list should be a good pair of close-focusing binoculars, a 10x hand lens (also known as a loupe) and an aerial net. With binoculars the key word for successful dragonflying is "close." Many companies now make models that can focus down to six feet or less. My pair easily focuses on my boot tops—while I'm wearing them! It is difficult to overstate how helpful a pair of good binocs can be. A full explanation of binocular features and a buyer's guide with specs on more than 38 models are found in Appendix E.

The author's daughter Christina nets early season odes. Learning to net and properly handle living dragonflies and damselflies can open up a whole new world of identification and appreciation.

If you would like to identify the species of most of the odonates you see, then sooner or later you'll want to get an aerial net and a hand lens. The reason is simple—to identify many species, you must examine small body parts under magnification. Close-focusing binoculars will help, but won't always do the trick. Netting is the best way to get odonates into your possession for a closer look. Look for a net with a deep bag, a rim with an opening of at least 15 inches in diameter and a lightweight handle at least 2.5 feet long. Some nets come with extendable handles, allowing you to add segments to get any length you want. Others are collapsible for easy travel. Bear in mind that a longer, heavier net is harder to swing quickly, and swinging quickly is really the key. Children do best with a smaller net, with a 10 to 12 inch diameter opening. Some enthusiasts use much smaller nets with success, like the kid's bug nets stocked by larger department stores. A good friend uses a really cute one (has painted eyes on the handle), which he sticks in a back pocket. Larger nets are more versatile and give you a larger margin for error when swinging. Smaller nets are easier to carry, may look less geeky and might draw fewer spectators. You decide what's important to you! Nets can be homemade or purchased from a number of equipment sources. BioQuip Products stocks a variety of net sizes and styles and also sells replacement bags. Rose Entomology and Acorn Naturalists also sell aerial nets.

Larger dragonflies in flight should be swung at from behind or below because of their lightning quick evasiveness. Netting them head-on is usually futile and even if successful, runs a higher risk of injuring them. Elusive species may require you to set up a bit of an ambush for you to get close enough for a good swing. An underwear-ripping swing is not usually necessary—just a quick, deliberate stroke followed by a quick turn of the handle so the rim cuts off the insects' escape from the net bag. Damselflies and smaller dragonflies are usually not too difficult to net. If an odonate is perched high on a twig, swing the net through the odonate and its perch. Damselflies in flight are less evasive than dragonflies and can usually be netted from any direction with a reasonably quick stroke. Many odonates regularly perch on vegetation near the ground, so simply approach them quietly, and quickly drop your net over the top of them. Netted odonates will usually fly upward toward the tip of an inverted net. Keeping the net rim low, reach your hand under the rim and up into the cone and use your thumb and forefinger

Equipment Sources

BioQuip Products: A wide selection of aerial nets (including the popular collapsible net), aquatic nets, Odonata envelopes, field guides, scopes and other entomological equipment.
(310) 667-8800
www.bioquip.com

Rose Entomology: A wide selection of entomological equipment including premium aerial nets.
(877) 249-1623
www.roseentomology.com

Acorn Naturalists: A great source for aquatic dip nets, books and other naturalist and teaching resources.
(800) 422-8886
www.acornnaturalists.com

International Odonatological Research Institute: Best source for collecting envelopes. Also field guides and other books.
www.iodonata.net

to gently grasp the four wings of the odonate together. The closer to the base you hold the wings the less likely you are to break any of them. Now you can examine the specimen with a hand lens or prepare it to bring home to examine under a microscope. Very small species must be handled with considerable care to avoid injuring them, and tenerals should be handled only with discretion as they almost certainly will be damaged. Most mature odonates that are netted in the way just described can usually be released unharmed, although a small proportion will be injured.

Using a hand lens (loupe) takes a bit of getting used to, but if you'll make a tiny bit of effort, it will open up a whole new world for you. A loupe will allow you to identify many more dragonflies and damselflies than you could without one. The key is simply to hold the loupe up near your eye with one hand while moving the specimen held gently between your fingers in your other hand until the odonate, and body part you need to see, comes into focus. That's it—move the specimen, not the loupe, to focus. Practice on a leaf or anything else that's not moving first. I recommend starting with a 10x loupe because it will be easier to use than one with higher magnification. After you get good with your 10x loupe you can think about supplementing it with a 14x or 16x model.

I urge you to consider keeping field notes of your outings. Doing so will help you learn to identify and know the habits of the species in your area more quickly. Taking notes is really quite enjoyable and will add richness to your outings. From May through October, I rarely go anywhere without my field notebook. It's even in the car on Sunday mornings when I'm on the way to church! You just never know when you're going to see something interesting. The most important things to record along with the species seen are the dates and locations where you found them. Locations should be described as precisely as possible, because if you find a rare species, other people will want to know where to look. Other things you may want to record are time of day, weather conditions, behaviors and stages of maturity of the species you see, and evidence of breeding and habitat notes, including vegetation if you know something about it. If you keep up with your notes diligently, over time they will become increasingly valuable both to you and from a scientific perspective. People who work with dragonflies and damselflies professionally are quick to affirm that citizen enthusiasts have greatly increased scientific knowledge about species distributions, critical habitats and behaviors. Any small notebook will do, but many people prefer Rite-in-the-Rain-type notebooks because they have waterproof paper. Always use pencil or water-proof ink because many inks will run when wet.

Proper dress is important to an enjoyable outing, including sunscreen and bug juice when appropriate. I like to wear a hat with a wide brim, lightweight field clothing and breathable waders that are scarcely heavier than slacks. I use hip-high or chest-high waders depending on the water depth where I'm going. Some folks are perfectly happy just to wet-wade in tennis shoes—which is fine for them, just don't ask me to do it! Knee-high boots will suffice in many situations, but don't be surprised if you go over the tops. A canoe or other small boat can get you to deep or muck-bottomed areas that would be impossible or unpleasant to wade to, but be careful—swinging a net from a canoe is tricky business, unless you enjoy swimming.

Gaining access to good sites for dragonflying can require a bit of planning. If you want access to private property, be sure to ask permission of the landowner first. When in doubt, ask—it's better to be safe than sorry. Depending on ownership, different public properties have different rules about public access and about the activities that are permissible on the property. Property access requirements may vary from state to state—be sure you do your homework. If you'd like to collect specimens, you may need to obtain a collecting permit in advance.

Collecting

Although not for everyone, some odonate enthusiasts choose to maintain a collection of these insects. Many scientists who study them do so of necessity as well. The reasons for collecting specimens are many. Collected specimens, when properly preserved and stored, are permanent and verifiable records of species distributions. It is important to document these distributions because they can change over time in response to losses of habitat, influences of invasive species, changes in climate and other factors. Management agencies charged with the responsibility of conserving and protecting natural resources for future generations need to know what species occur in different areas and what habitats are critical to their survival. Many state and provincial agencies now use citizen volunteers in surveys to help determine distributions and critical habitats of species. Usually, some collecting is required to contribute to these surveys in a meaningful way.

Collections play an important role in the study of Odonata.

For some species, collection is necessary because careful examination under a microscope is the only way to identify them with certainty. Reference specimens are crucial for scientists who study identification and the phylogenetic relationships of species. Specimens are needed to properly determine relationships of closely related species and groups. Besides being essential to research, collecting specimens can be a source of pleasure and education. Responsible collecting poses no threat to odonate populations, although restrictions on collecting are appropriate when populations are dangerously low, habitats are especially fragile or legal reasons apply.

In sum, collecting is a great way to learn, and it is a crucial tool in the protection and conservation of odonates. If you are thinking about starting a collection, you should first read the Responsible Collecting Guidelines put out by the World Dragonfly Association and the Dragonfly Society of the Americas on their websites (see Appendix B). Then learn the proper techniques for collecting, preserving and storing specimens, either from these websites, or from a reliable, up-to-date odonatological text such as Paulson (2009). When you're finished with it, a properly curated collection can be donated to a museum or university.

Physiographic Regions

Physiographic regions are large-scale subdivisions of the United States based on terrain texture, rock type and geologic structure and history. Within the U.S. Rocky Mountains there are three physiographic regions: the **Northern Rockies**, the Middle or **Central Rockies** and the **Southern Rockies**. There is not universal agreement among scientists as to exactly where the boundaries of these regions should be placed. Some people combine the Northern Rockies and Central Rockies into one region. For the purposes of this guide, the map on page 36 defines the regions.

The Northern Rockies includes much of central and northern Idaho, and most of western Montana. This region includes the Cabinet, Clearwater, Salish and Salmon River Mountains, and the Bitterroot and Sawtooth Ranges. Glacier National Park is located here and hosts the highest peaks in the region, several of which are over 12,500 feet. The Central Rockies includes parts of southeastern Idaho, northeastern Utah, and southern Montana, and much of western Wyoming. The region includes the Bighorn, Beartooth and Uinta Mountains, and the Absaroka, Gallatin, Snake River, Teton, Wasatch and Wind River Ranges. The region includes Grand Teton and Yellowstone National Parks and several peaks in excess of 13,500 feet. The Southern Rockies form a broad region that extends from north-central New Mexico north of the Rio Grande well into southeast Wyoming at the Great Divide. It includes most of the western 2/3rds of Colorado, and Rocky Mountain National Park. The region is formed by two north/south belts of mountains separated by three basins known as the North, South and Middle "parks." The eastern belt boasts the Laramie, Medicine Bow and Wet Mountains, and the Front Range. The Western belt includes the Sangre de Cristo and San Juan Mountains, and the Gore, Park and Sawatch Ranges. The Southern Rockies include the highest mountains within the U.S. Rocky Mountains, with more than 50 peaks over 14,000 feet.

In the Rocky Mountains we get a great blend of eastern, western, northern and southern species, as well as species that occur throughout the continent. Considerable overlapping of eastern and western species occurs, but eastern species are most common or even entirely restricted to the east of the Continental Divide, and likewise, most western species are found in greatest number to the west of it. We are fortunate that many boreal species extend their ranges southward at high elevations throughout the Rockies. And some southern or plains species are found in the Southern and Central Rockies, mostly in the foothills, but sometimes creeping up into the lower Montane Zone. Few regions of North America can lay claim to as generous a mix of eastern, western, boreal and plains species.

Biotic Zones

The mountain environment is usually divided into three main biotic zones (a.k.a life zones): the **Alpine**, **Subalpine** and **Montane**. These regions are based on mean summer temperatures and therefore do not occur at any fixed elevation; rather, they occur at increasingly higher elevations as latitude decreases (the further south you travel in the

Biotic Zones

	Approx. elevation in Colorado (ft.)	Approx. elevation in Montana (ft.)
Alpine	12,000 – up	7,500 - up
Subalpine	9,000 – 12,000	6,500 – 7,500
Montane	6,000 – 9,000	4,000 – 6,500
Foothills	5,500 – 6,000	3,500 – 4,000

Rockies). The chart above shows the approximate elevations of the zones in Colorado and Montana. In Wyoming the elevations of the zones would be roughly intermediate between the two.

Most Biologists agree that the **Alpine Zone** begins at and ascends from the point where the average temperature for the warmest month of the year is about 50 degrees F. This is the zone above the tree line with an open, but highly diverse landscape of heather, willows, grasses and wildflowers (called **tundra**). Most aquatic habitats in this zone are ponds and small lakes that vary greatly in hydrology. A small handful of adult odonate species have been found (rarely) in the lower parts of the

The Rocky Mountains can be divided into Northern, Central and Southern regions.

Alpine Zone, but if any odonate breeding occurs here, it is limited to the lowest part of it. I well remember my first visit to an alpine wetland and the lesson I learned about the harshness of this environment. It was August 1st and I came over a crest to see a beautiful depressional wetland shimmering in the sunlight below me. Excitedly, I ran down a slight slope to the wetland, aerial net in hand, only to find it covered with ¼ inch of ice!

The **Subalpine Zone** is the biotic zone immediately below the Alpine Zone and below the tree line. The zone is characterized by Subalpine Fir and Engelmann Spruce, with a few pines and other firs occurring in some areas. The climate is cool with summer high temperatures on the warmest days averaging in the upper 60s (degrees F) in the middle of the zone and 75 degrees F only near the Montane Zone. Snowdrifts here may linger into summer. This is the highest zone with consistent odonate activity. High elevation species breed in ponds, lakes and other still-water habitats throughout the Subalpine Zone, but streams here typically produce odonates only in areas of reduced flow, like beaver impoundments, at least in the upper parts of the zone.

The **Montane Zone** is the mountainous zone with the greatest odonate diversity and the greatest diversity of aquatic habitats. Located between the Subalpine Zone and the Foothills, the Montane Zone forms the lowest forest belt in the Rockies. Most common tree types here are Ponderosa Pine, Lodgepole Pine, Douglas Fir and Quaking Aspen. Streams and rivers of all sizes abound. Larger rivers are usually regulated (with dams, diversions or abstraction for water supply purposes). Still-water habitats run the full gamut of lakes, ponds, marshes, swamps, bogs and fens.

Below the mountains lie the **Foothills** which are typically dominated by a variety of dry land shrubs, grasslands, relatively open woodlands. It is beyond the scope of this guide to specifically address odonate species that thrive at lower elevations, other than to say that many Foothills species have ranges that extend up into the Montane Zone.

Species Included in this Field Guide

A complete list of Odonata species known to breed in the six-state Rocky Mountain region does not exist. Species lists are usually available for areas with clearly defined boundaries such as states or National Parks. Because the Rocky Mountain regions do not have such firmly defined boundaries, I settled upon a reasonable mapping scheme for the three regions (see previous page) and included those species that have been reported at least several times within one or more of those boundaries. The focus is the U.S. Rockies south of Canada, so species that occur only in the Canadian Rockies or the Brooks Range in Alaska are not included. To make decisions about species to include, I consulted a vast array of published and online resources, the most important of which were Donnelly (2004 a, b, & c), Paulson (2009), and the distribution resources at Odonata Central (www.odonatacentral.org). To keep this field guide to a manageable size, a small handful of species that were only represented by one or two specimens, sometimes from many decades ago, were not included. A couple of these reported species were well outside of their established ranges and their occurrence in the Rockies could not be validated with a specimen. With more reporting being done all the time, any of these excluded species could turn out to occur more regularly in the Rocky Mountains than currently thought. However, the odonates you will see during your trips to the mountains are very likely to be found among the 99 species in this guide, and you won't have to thumb through extra pages of unnecessary species to find them.

How to use this Field Guide

Dragonflies and Damselflies of the Rocky Mountains is designed to make field identification easier for you, the reader. Through the use of color photos, arrows pointing to field marks, size scales, phenograms and habitats, we have made a handy, compact and easy to use guide. Also, by limiting the dragonflies and damselflies to those found in one geographic area, we have eliminated the need to wade through several hundred species, many of which would never be found here. Included is every regularly occurring odonate in the Northern, Central and Southern Rocky Mountains of the United States.

Order

Dragonflies and damselflies are organized by families and then broken down further into genera. Family name, ending in idae (pronounced *idée*), is listed at the bottom of the left page of each species spread while the genus is listed on the bottom of the right page. We generally placed species in their currently interpreted phylogenetic order, but placed similar-looking species next to each other regardless of phylogeny. After all, this is first and foremost a guide to identification. With experience in the field using this guide, you will gradually learn to identify odonates to family, then genus and then species.

Dragonfly and Damselfly Names

Like other organisms with popular appeal or economic importance, dragonflies and damselflies are given common names as well as their obligatory scientific names. Common names are the English names most naturalists use. Scientific names, which have Greek or Latin roots, tend to be the spoken word of odonatists. Species are described scientifically by their capitalized genus name, followed by their lower case species name. We capitalize species common names but use lower case for groups (e.g. River Bluets are actually a type of bluet).

Photos

Primarily we use photos of free-flying dragonflies and damselflies in their natural habitat. Males are pictured on the left page of the spread and females are pictured on the right. The symbol used for males is ♂ and the female symbol is ♀. Photo credits are listed in Appendix C.

Fieldmark Arrows

Arrows point to diagnostic features in the photos that are referenced in the description text and marked with an arrow symbol (↑). These are characteristics that you should look for while in the field. Jotting down notes on size, color, thoracic stripes, abdomen markings, reproductive parts and wing features will help you identify the odonate when you have a chance to consult this book.

Size Scale

Size is relative and often hard to judge in the field. Use the size-bars at the top of each species' main photo. The black bar indicates the actual body length of that species.

Phenograms

All odonates live out their lives according to seasonal timing that is characteristic for that species. Our phenogram highlights in red the time when that species is active as a flying adult. In other words, look for that odonate during the highlighted weeks/months. Of course, weather, elevation and latitude will cause variation in emergence dates and flight period. As a general rule, species at lower elevations and more southern areas will become active during the earlier part of the phenogram, whereas species at high elevations or more northern areas will become active during the later part of the phenogram and may be active for a shorter period of time.

Breeding Habitat

Preferred breeding habitat is found beneath the phenogram. Remember, this is where courtship, mating and egg laying take place, and the nymph develops and emerges. Your best bet to find the species will probably be near such habitats—but not exclusively. Damselflies and especially dragonflies can range widely over many different habitats.

Nature Notes

Nature Notes are fascinating bits of natural history that bring one a more complete understanding of that species. Unique behavior, population trends and naming history are just some of the topics touched on.

Range Map

A map of the U.S. Rocky Mountains shows the range of that species in orange. This is the range as we know it from current population data, but some odonate species distributions are not yet fully understood.

Species Text

Description covers the best distinguishing characteristics—whether it is the abdomen markings, shoulder stripes, wing color, reproductive parts, wing veins or eyespots. If males and females are different (sexual dimorphism) then these differences are described. Measurements given are the average body length in inches.

Under **Identification Clues**, tips are given on how to separate that species from other similar odonates. Details of spreadwing male claspers (terminal appendages) and female ovipositors are shown side-by-side on page 55. Details of pond damsel male claspers are shown on pages 76-77. Thoracic side stripes of male mosaic darners are provided on pages 146-147. Important reproductive parts of clubtails, emeralds, whitefaces and meadowhawks are illustrated on pages 177, 197, 217 and 257 respectively. These features usually require a hand lens to see clearly.

Life Cycle covers flight season, courtship, male guarding, egg laying technique and nymphal development.

Rocky Mountains Odonate Checklist

In Appendix A you will find a checklist of all 99 species found in the U.S. Rocky Mountains of Idaho, Montana, Wyoming, Utah, Colorado and New Mexico. Check off the ones you see in your travels afield. We've also included checkboxes specific to the four largest national parks in the U.S. Rockies: Glacier, Rocky Mountain and Yellowstone/Grand Teton.

Enjoy *Dragonflies and Damselflies of the Rocky Mountains.* Take it with on hikes. Stuff it in your backpack. Use it. But most importantly, have fun getting to know our fascinating Rocky Mountain odonates.

Identification

Most Rocky Mountain odonates have readily observed markings or structural characteristics such that with the help of close-focusing binoculars or even the naked eye, you will learn to identify free-flying individuals without capturing them. However, some groups of species are so similar to each other in appearance that identification requires capturing them and examining small or awkwardly placed body parts closely in the hand, usually under magnification. Here are a couple of basic principles to remember. First, being able to determine the gender is almost always necessary (see page 9). Second, males are usually easier to identify than females. Of course, there are exceptions. For example, because of her outrageously long ovipositor, the female Sweetflag Spreadwing is easier to identify than the male. Third, mature adults are easier to identify than tenerals. As young individuals mature, their body colors and their eye color will usually change. Bear in mind also that circumstances occur where even specialists may not attempt to identify every odonate they see, especially the teneral females of some pond damsels and meadowhawks. This is because the uncertainty of the result, and the risk of damaging the specimen, may make the effort less than worthwhile. Females of difficult genera can often be tentatively identified by identifying the males they are associating with.

A 10x loupe is essential gear for the serious student of odonates.

Three levels of identification are typically used with odonates: in-the-field, in-the-hand and under-a-microscope. **In-the-field** refers to species that can be identified by field markings with the naked eye or close-focusing binoculars in their natural state. **In-the-hand** refers to species that must be captured and examined more closely because the body parts or marks used for identification are too small to be seen at a distance, or must be viewed at a particular angle. For this level of identification I normally use a 10x hand lens. **Under-a-microscope** refers to identifications that can only be reliably made with magnification of greater than 20x. This is particularly true of some female pond damsels.

Dragonflies and Damselflies of the Rocky Mountains will enable you to identify the vast majority of Rocky Mountain odonates in-the-field and in-the-hand. This level of success represents what I believe is reasonably achievable for most nature enthusiasts. The under-a-microscope level is not addressed in this guide. As your interest in odonates deepens, you may at some point want to invest in a decent stereomicroscope, light source and appropriate dichotomous keys. But you can cross that river when, and if, you come to it.

To identify an odonate, first use the Family Keys and descriptions to determine the family of your subject (pages 44 and 45). Then compare the body size, color patterns, clasper shapes, eye size and color, wing markings and key aspects of wing venation, with the photos and illustrations in this guide. Among the damselflies, the broad-winged damsels are easily identified in-the-field by their size, coloration and wing markings. For spreadwings and pond damsels, capture some males with your net and examine their abdomen tips (terminal appendages) with a hand lens, comparing them with the illustrations in this guide. Also notice the patterns of stripes, spots and rings. First determine if the specimen is a spreadwing or a pond damsel using the Family Key on page 44. Spreadwings hold their wings partially open at rest, are usually larger than pond damsels, and often are pruinose at the tip of the abdomen. Once you have your choice narrowed down to a group of species, use the identification clues given in the species accounts to rule out similar species. In this way, you can use this guide to identify mature males of virtually all Rocky Mountain damselflies and more than half of the females you will see.

Among the dragonflies, some of the darners can be identified in the field, when perched, if you can get a good enough look at their thoracic colors and side stripes, and the claspers at the abdomen tip of the males (pp. 146-147). However, you will need to net most darners and use the identification tips on the species pages to identify them. Our few clubtails are similarly best identified in-the-hand with a hand lens because you will need to see their small reproductive parts close up (page 177). Our single species of spiketail and cruiser are both readily identified in the field by their large size, coloration and thoracic side stripes. Emeralds should be netted and examined in hand until you become familiar with the various species because they are mostly brown and roughly the same size. Like clubtails, emeralds are best identified by the shapes of small reproductive parts (page 197). Our skimmers are diverse in size and color, and most perch frequently and are easy to approach. Therefore, in-the-field determinations are often possible, especially with the whitetails, corporals, king skimmers, rainpool gliders and saddlebags. Whitefaces and meadowhawks can be tricky to identify, so get in the habit of netting some and comparing their small reproductive parts with the drawings on pages 217 and 257.

I want to leave you with a final identification thought: it never hurts to net an odonate and examine its key structures closely even if you think you know what it is. Sometimes you will be surprised. Remember, color patterns within any given species can be vexingly variable, but structural parts such as hamular processes, claspers and subgenital plates, generally are not. So, if you really need a positive identification, you should strongly consider capturing the specimen. As you get increasingly familiar with more and more species, there will be less need for netting. But netting and closely examining odonates with a loupe is a great habitat to get into when starting out, and it is a habit that most experts stick with.

Family Key for adult Rocky Mountain Damselflies

1a. Wings colored with black, brown or red, with no "stalk" at the base and with many antenodal cross veins (see drawing below); body metallic green or bronze; large species: BROAD-WINGED DAMSEL of FAMILY CALOPTERYGIDAE (pg. 46)

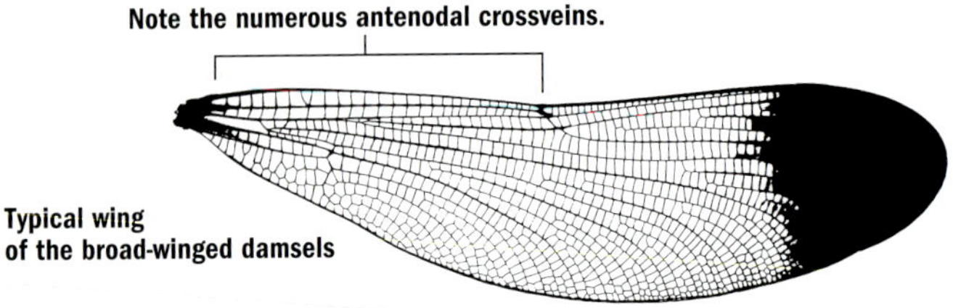

1b. Wings clear or lightly colored with amber, with a distinct "stalk" at the base and with just two antenodal cross veins; body may be any color; large and small species: go to 2

2a. Wings held partially open at rest unless teneral; wing median vein splits closer to the arculus than the nodus (see drawing below); mostly larger species: SPREADWING of FAMILY LESTIDAE (pg. 52)

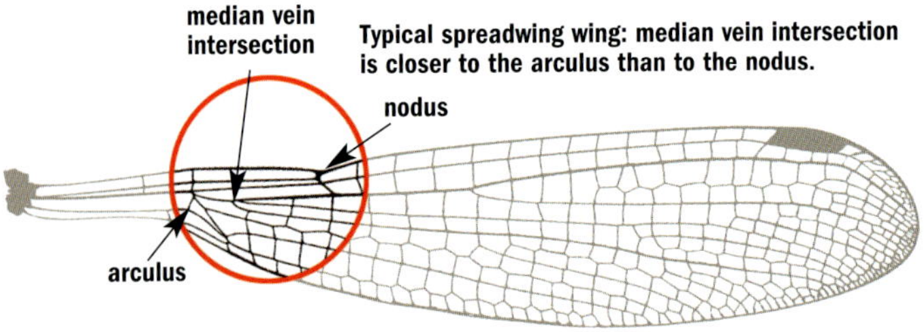

2b. Wings held together against or over the abdomen at rest (except Taiga Bluet); wing median vein splits closer to the nodus than the arculus (see drawing below); mostly smaller species: POND DAMSEL of FAMILY COENAGRIONIDAE (pg. 68)

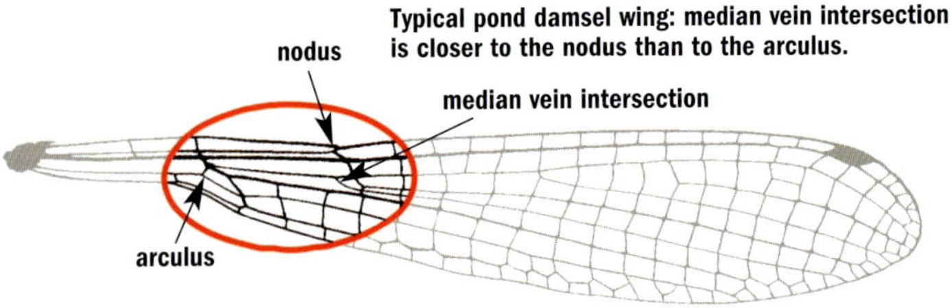

Family Key for adult Rocky Mountain Dragonflies

1a. Eyes well separated, not touching on top of head; small to mid-size dragonflies with clear wings and abdomen with small to pronounced club (wider segments near tip): CLUBTAIL of FAMILY GOMPHIDAE (pg. 174)

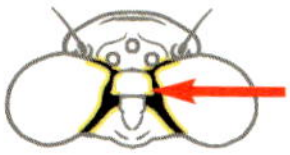

1b. Eyes meet broadly along a seam on top of head, or just at a point; abdomen not clubbed; body size and wings variable: go to 2

2a. Eyes meet at a single point of top of head or barely separated; eyes green or blue; body color dark brown or black; thorax with 2 conspicuous yellow side stripes; wings clear; large species: SPIKETAIL of FAMILY CORDULEGASTRIDAE (pg. 188)

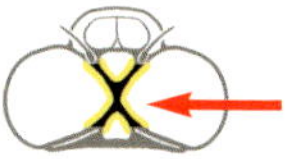

2b. Eyes meet along a distinct seam on top of head; other characteristics variable: go to 3

3a. Single yellow side stripe on thorax; single row of large pale spots down abdomen; body brownish gray; dark stripe across face; eyes gray and meet along seam on top of head for short distance; large species: CRUISER of FAMILY MACROMIIDAE (pg. 189)

3b. Thorax lacking single pale side stripe (may have two pale stripes, one or more spots, or be unmarked); other characteristics variable: go to 4

4a. Eyes very large, covering much of head, and meeting along a long seam of top of head; males with brightly colored bodies with blue, green or yellow markings; long slender abdomen; large, strong-flying species that tend to perch vertically: DARNER of FAMILY AESHNIDAE (pg. 142)

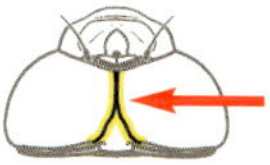

4b. Eyes moderate in size, meeting along a short seam on top of head; other characteristics variable: go to 5

5a. Mid-sized, light to dark brown species; body often with a shiny or metallic luster; wings clear; eyes often brilliant green in mature adults; anal loop of hind wing elongated, but without distinct boot shape and lacking a pointed "toe;" perches vertically or obliquely: EMERALD of FAMILY CORDULIIDAE (pg. 194)

5b. Small to large species with wide range of colors and patterns; wings clear or obviously patterned; anal loop of hind wing elongated and with distinct boot shape and pointed "toe;" perches horizontally or obliquely: SKIMMER of FAMILY LIBELLULIDAE (pg. 214)

Broad-winged Damselflies Family Calopterygidae

The broad-winged damsels are a small group of large, showy species that exclusively inhabit the flowing waters of rivers and streams. They are readily separated from our other families of damselflies by easily seen characteristics of the wing, including color pattern, shape and wing venation. The family name, Calopterygidae, means "beautiful wing," referring to the striking patterns of wing coloration in both of our Rocky Mountain species. Wing shape in this family differs from other families in that the wings are broad and lack a distinctly narrowed "stalk" at the base. Among the many wing venation characteristics peculiar to the group is the presence of many antenodal crossveins on each wing, whereas our other families only have two antenodal crossveins per wing.

Males of our two broad-winged species engage in showy courtship displays and defend small territories along the stream's edge. Neither species ventures far from water. At the breeding sites, females often match males in numbers. Females exercise choice in whether to accept or reject a displaying male. Eggs are deposited into plant stems that are often well below the water's surface. Males do not remain in tandem when eggs are laid, but rather, guard in non-contact fashion by hovering or perching nearby.

The nymphal stage lasts for at least a year and probably two or even three years for the River Jewelwing. The nymphs crawl about on submerged roots, plant stems and other forms of vegetation along the edges of flowing waters. Not surprisingly, the nymphs are large for damselflies and have broad wing pads.

About 180 species of broad-winged damsels are known worldwide (mostly tropical), with only eight species found in North America north of Mexico. Here in the Rocky Mountain Region we have one species in each of two genera. Both sexes of both of our species are easily identified in the field by wing shape and color patterns.

The Jewelwings — Genus *Calopteryx*

The jewelwings are so named for the many species that have iridescent wings. Our single species has wings that are partially black and the females show bright white patches (pseudostigmas) at the tips of their wings. Both genders lack true stigmas. The legs are long, slender, dark and adorned with long spurs.

Male jewelwings vigorously defend their "turf" along the edge of a stream, chasing away other males. In their territories they perform elaborate courtship displays, featuring wing-flap displays and other special wing movements (see photo on pg. 48). Males will also alight on the surface of the water near the egg-laying site in what is called a dive display. This demonstrates to the female both the quality and location of the oviposition site. If a female is receptive to mate, she may allow the male to land on her closed wings, at which time he walks up along the wing's leading edge before taking her in tandem. To reject the male's display, the female simply spreads her wings and curves her abdomen up. The flight of jewelwings is usually described as bouncy and butterfly-like.

The Rubyspots — Genus *Hetaerina*

The rubyspots get their name from the unmistakable, blood red patches at the bases of the wings of males. The scientific name means "little companion," which Dennis Paulson and Sid Dunkle suggest may refer to the red armbands of Greek courtesans. The wing venation pattern is similar to that of jewelwings, but the wings are a bit narrower, have a small stigma and lack extensive areas of black. Males will defend their favorite perches from other males using a circle flight pattern. Female American Rubyspots will also defend territories; territorial females are unusual among odonates.

Jewelwings (Genus *Calopteryx***)**

Rubyspots (Genus *Hetaerina***)**

River Jewelwing *Calopteryx aequabilis*

APRIL	MAY	JUNE	JULY	AUG	SEPT	OCT

Breeding habitat: Clean streams, medium-sized or larger, with open canopy in Foothills to lower Montane zones. Also small to medium-sized rivers.

Nature Notes:

Species name means "equal," perhaps referring to the nearly half black hindwing.

Male courts female with display of slow alternating wing beats (see photo below). Male's courtship may include flinging himself onto the water's surface as if to show the female that the current speed in his territory is ideal.

Description: Adults average 2 inches long.

Head, thorax and abdomen bright metallic green, and may appear blue in some light. Male's wings clear at base, banded with black at the tip (↑). Hindwing's black band is larger ($^1/_3$ of wing) than forewing band ($^1/_4$ of wing). Female's wings are light brown to smoky and usually darker at the tip. Her dark wingtips show less contrast with the rest of the wing than in the male. The white pseudostigma near wing tip of female is conspicuous (↑). Wings of both sexes 3 $^1/_2$- to 4-times longer than wide.

Identification Clues: Both genders are unmistakable. No other Rocky Mountain damselfly male has a black patch at the wing tip. Female is unique among Rocky Mountain species in being large in size, having solid

Although the female River Jewelwing's entire wing is often dark, the outer third is usually blacker than the rest.

metallic green thorax and abdomen and having dark (brown) wings with prominent white pseudostigmas.

Distribution: All Rocky Mountain regions in Foothills to lower Montane Zone.

Life Cycle & Behavior: Adults emerge in last half of June or July and may fly into September. Eggs are usually laid into submerged plant stems. Female may crawl a foot or more below the water's surface to lay eggs and stay submerged for 40 minutes or more. Nymphs progress through 12 or 13 molts and probably require two or three years to mature.

American Rubyspot *Hetaerina americana*

APRIL	MAY	JUNE	JULY	AUG	SEPT	OCT

Breeding habitat: Gently to rapidly flowing larger streams and rivers (> 15 feet wide) in Foothills and lower Montane regions.

Nature Notes:

May congregate in densely packed groups. One Ohio odonatist described catching 75 with one sweep of his net!

Perches on vegetation over water or on rocks.

May obelisk, pointing abdomen tip toward sun, at high temps.

Flits over shallow riffles where stream banks are well-vegetated. Does not wander far from stream.

Late season species that often flies through August.

Female may descend 60 cm underwater to lay eggs.

Description: Adults average 1.7 inches long.

Male's head and thorax are red to reddish brown with thin pale lines along the sutures that become obscured with age. Wings are clear with a blood-red patch infused with white veins at the base (↑). Abdomen is metallic bronze to dark brown with pale, narrow rings on middle segments. Female's head and thorax variable—metallic reddish-brown, bronze or green with thin, pale lines along the sutures. Female's wings range from mostly clear to having yellowish or light brown wash, often with reddish or amber patch at base (↑). Abdomen shades from metallic green to bronze to dark brown in mature males, with pale, narrow rings on middle segments.

Identification Clues: Blood-red patches at base of male's wings are unmistakable—only

Metallic green thorax is characteristic of mature females like this one.

one rubyspot species in Rockies. Female easily identified by large size, heavily veined wings, pale-colored stigma, pale sutures through metallic green or bronze thorax, reddish suffusion near wing base and abdomen with pale rings.

Distribution: All Rocky Mountain regions in Foothills and lower Montane Zone.

Life Cycle & Behavior: Adults emerge in late June or July and may fly well into September. Eggs are laid into plant stems, willow roots or into partially submerged, rotted wood, usually just a few inches underwater. Female may stay submerged for nearly an hour. The nymphal stage has not been extensively studied. Nymphs likely overwinter in the 2nd to 4th instars and are ready to emerge in one year.

Spreadwings Family Lestidae

The spreadwings differ from all other Rocky Mountain damselflies in that they perch with their wings held partially open when active, at about a 45 degree angle to the body (hence their common name). The only other adult damselfly in our area that habitually perches with its wings slightly spread is the Taiga Bluet, which is easy to separate from the spreadwings by its color pattern and smaller size. A word of caution—don't use the open-winged position to identify tenerals, because they refuse to play by the rules. Teneral pond damsels often perch with their wings spread, like spreadwings, and newly emerged spreadwings hold their wings together over their backs like pond damsels.

Spreadwings are medium to large in size and are usually found at still-water habitats including temporary ponds and vernal pools. They like to perch on vertical plant stems holding their bodies downward, usually at an oblique angle. Spreadwings are usually not brightly colored, although the bodies of some are a wonderful iridescent green or bronze. Most are gray, brown or black, with various amounts of pruinosity. These drab colors blend in well with the rushes, reeds and sedges of their preferred haunts. Mature males often have bright blue eyes and faces, and the tips of the abdomens of most species appear white or grayish-white because of pruinosity. Males tend to become increasingly pruinose with age, often obscuring the underlying colors and pale stripes of the thorax. The wings of spreadwings are narrow, distinctly stalked at the base (called petiolate), not heavily veined, generally clear and have long stigmas. The wing venation of the group is diagnostic because the position on the wing where the median vein splits sets them apart from pond damsels (see pg. 44), as does their long stigmas.

Approximately 151 species of Lestidae are known worldwide, of which 19 are found in North America north of Mexico. Six species, five in the genus *Lestes* and one in the genus *Archilestes*, are found in the Rocky Mountains. All males can be identified in the hand by the shapes of their terminal appendages (claspers), in conjunction with other characters. Some species (both sexes) can be identified in the field because of their unique body shape, body coloration or elongate ovipositor. However, females of a few species can be identified with certainty only under a microscope, primarily by subtle differences in the lengths of some abdominal segments and color on the rear of the head.

Stream Spreadwings — Genus *Archilestes*

These are very large damselflies, the largest in North America, and they look considerably larger than any western pond spreadwing or pond damsel. Prominent yellow stripes on the sides of the thorax are easily visible, not becoming obscured by pruinosity, which develops only on the abdomen tip of the male. Males defend small territories. They are primarily stream species but are often found at ponds as well. Females are unusual in laying their eggs into woody stems that may be far above water.

Pond Spreadwings — Genus *Lestes*

The pond spreadwings are just about impossible to confuse with any other group. No other medium-sized damselflies perch on vertical stems with abdomen lowered and wings held apart. Also, the blue eyes and grayish-white (pruinose) abdomen tips of the males are unmistakable. However, pruinosity increases with age for both males and females, which can complicate species identification. If you take a close look at one, notice that the spurs on the lower legs are unusually long—much longer than the spaces between them. Many aspects of the pond spreadwing life cycle are also unique. As with the Great Spreadwing, all of our pond spreadwing species spend the winter in the egg stage, which is unique among Rocky Mountain damselflies. This egg-wintering strategy allows them to live across the entire spectrum of still, freshwater habitats including temporary ponds that dry up during summer. How do they do that?

All of our damselflies, other than spreadwings, lay their eggs into plant stems near or below the water surface where moisture and appropriate temperatures quickly start the egg hatching process. These species all need permanent waters because the eggs hatch before winter and the nymphs must spend the winter underwater, which is typical of the vast majority of odonates. However, our female pond spreadwings lay eggs, in tandem with the male, into the stems of cattails, bulrushes or spikerushes often a foot or more above the water. The eggs immediately take some of the initial developing steps, but before hatching occurs they slip into a state of embryonic diapause (rest). In this state, the drying-resistant eggs spend the winter inside now-dead plant stems where they can withstand temperatures as cold as it gets below the Alpine Zone (with much needed help from an insulating cover of snow). Then, after the snows melt and water levels rise in the spring, the dead plant stems fall into the water (if they haven't already done so), and are wetted, providing the moisture needed along with the right increasing temperature and day-length cues to complete the hatching process. After hatching, the pronymph emerges from the cut in the stem, falls into the water and quickly molts into the second instar. Nymphal life is short for these egg-diapausing species; just two or three months in spring and summer. This is because the fast-growing nymphs need to complete their development of about ten instars and emerge before the temporary ponds they live in dry up.

Even pond spreadwing nymphs are unique. They are long, slender and agile, with long legs and large, leaf-like caudal lamellae. Flip one on its back and you will see an amazingly long labium (face mask) that is strongly narrowed at its base. No other Rocky Mountain damselfly has a labium quite like it. The nymphs cling to submerged vegetation where they patiently stalk their prey.

Pond spreadwings are usually found along cattail-, rush-, sedge- and reed-bordered shores of marshes, temporary and permanent ponds and small, sheltered lakes. A few species are infrequently found on slow sections of rivers and streams. Most of our species do best in waters with few or no fish, where the top predators are usually dragonfly nymphs in the darner family. The exception, the Emerald Spreadwing, does best in vernal pools or other waters that lack both fish and darner predators. Wherever they live, adult pond spreadwings flit from stem to stem along the shore in short, low, flights, feeding primarily on small flies that have emerged from the same areas. Territoriality and courtship behavior are absent or poorly developed in this genus. The name *Lestes* means "robber" or "pirate," probably referring to the predatory habits of adults or nymphs.

Stream Spreadwings (Genus *Archilestes*)

Pond Spreadwings (Genus *Lestes*)

Top view of male claspers

Side view of female abdomen tip

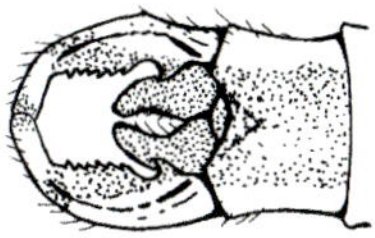

Spotted Spreadwing
Lestes congener

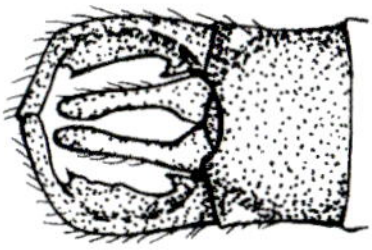

Sweetflag Spreadwing
L. forcipatus

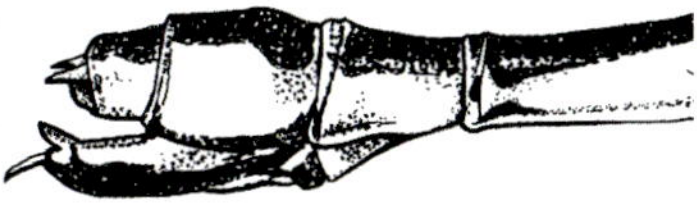

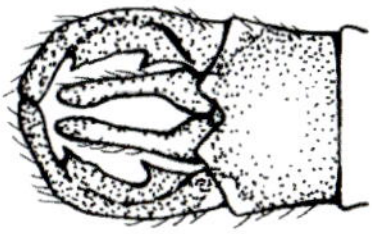

Northern Spreadwing
L. disjunctus

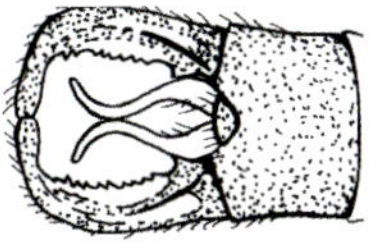

Lyre-tipped Spreadwing
L. unguiculatus

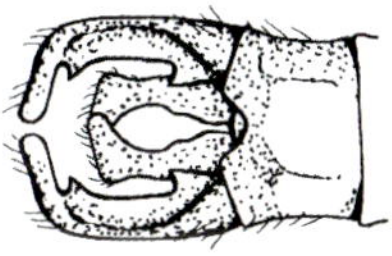

Emerald Spreadwing
L. dryas

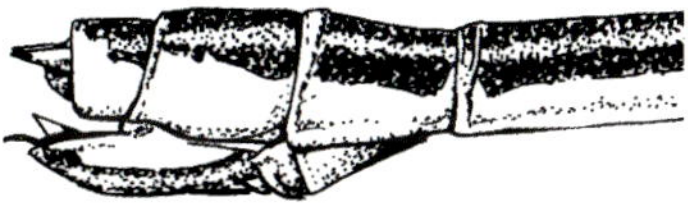

Great Spreadwing *Archilestes grandis*

APRIL	MAY	JUNE	JULY	AUG	SEPT	OCT

Breeding habitat: Slow streams, backwaters, ditches and some ponds in forested areas in Foothills to lower Montane regions of Southern Rockies.

Nature Notes:

Largest North American damselfly

Has dramatically expanded its range across the nation from the southwestern U.S. since the 1920s.

Female has been observed to lay eggs in a tree 44 feet above water!

The robust nymphs swim actively like tiny fish.

Description: Adults average 2.3 inches long.

Male has blue eyes, powder blue labrum and dark brown thorax with thin, metallic green shoulder stripes and two bold, yellow stripes on side of thorax. Thorax shows light coat of pruinosity as male matures. Male's abdomen dark brown or black with gray pruinose abdomen tip on segments 9 and 10. Wings clear with several unique features that differ from pond spreadwings. Male's claspers feature very short paraprocts, not easily visible is dorsal view. Female marked much like male but lacks pruinosity. Her eyes may be blue or brown and abdomen has pronounced bulb-like tip. Ventral margin of her ovipositor has large, coarse "teeth."

Identification Clues: Both sexes are unmistakable because of large size, prominent yellow stripes on side of thorax (↑) and widely out-

Bold yellow stripes on the sides of her thorax instantly identify this female.

spread wings (↑). Looks enormous compared to other damselflies!

Distribution: Foothills to lower Montane Zone of Southern Rockies.

Life Cycle & Behavior: Late season species in Rockies; adults emerge during summer and fly into September or later. Female lays eggs in tandem with male into herbaceous or woody stems or leaf petioles often 3 feet or more above water. One pair oviposited in tandem for an hour then the male left and the female continued alone for another 30 minutes. Egg overwintering is the rule in the north, but nymphs may overwinter in the south. Nymphs quickly complete development and mature in one growing season. Prefers fishless waters where nymphs swim actively in absence of fish predation. Species is known to tolerate poor water quality, which may also correlate with reduced numbers of predators and competitors at some sites.

Spotted Spreadwing *Lestes congener*

APRIL	MAY	JUNE	JULY	AUG	SEPT	OCT

Breeding habitat: Marshy and bog-bordered ponds, swamps, marshes and slow streams. Thrives in fishless habitats. Very tolerant of saline waters.

Nature Notes:

Our latest damselfly, often seen into September or October depending on latitude and elevation.

Huge numbers of pairs may simultaneously oviposit in rush beds.

Common and widespread throughout the Rocky Mountains, but you might miss it if not looking late in the season.

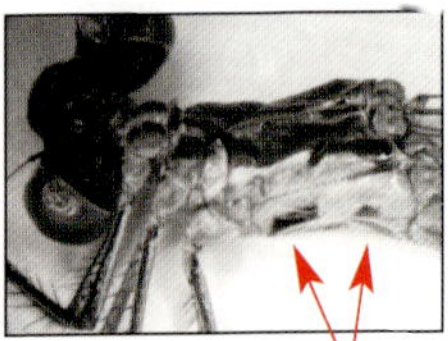

Look for the paired spots under the thorax.

Description: Adults average 1.4 inches long.

A small, dark and stocky pond spreadwing. Thorax and abdomen of male is dark above changing abruptly to pale whitish-gray sides and a pair of conspicuous dark spots on lower edge of thorax behind rear legs (↑). Thorax has very thin, pale shoulder stripes. Pruinosity on the thorax becomes more extensive with age. Abdomen is dark brown above, lighter on sides, with last two segments grayish-white pruinose. Wings clear. Male cerci about twice as long as the short paraprocts (page 55). Female is similar to male but with a thicker abdomen. Female ovipositor is short, not extending beyond tip of abdomen.

Identification Clues: Both sexes are easy to identify by the presence of a pair of dark spots on the lower edge of the thorax. Sweetflag Spreadwing often has only one dark spot there

Female is darkly colored, much like the male.

and Sweetflag is larger and has longer, thinner paraprocts. Small size for a pond spreadwing, dark coloration of thorax above contrasting sharply with pale sides and late flight period are in-the-field clues.

Distribution: Widespread through all Rocky Mountain regions and zones except for Alpine.

Life Cycle & Behavior: Adults begin to emerge in July and peak abundance is reached in August and September. Some fly well into October. Pairs form away from water then fly close to the egg-laying areas. Eggs are laid in tandem into dry bulrush, spikerush or cattail stems above water level. Eggs overwinter, but in an earlier state of embryonic development than other spreadwings. The small eggs are highly resistant to desiccation and extreme cold, surviving air temperatures at least as low as minus 18 degrees F, colder when snow covered. Eggs hatch in late spring. Nymphal development through ten instars takes about 50 days. Newly emerged adults require three weeks to mature.

Sweetflag Spreadwing *Lestes forcipatus*

APRIL	MAY	JUNE	JULY	AUG	SEPT	OCT

Breeding habitat: Well-vegetated or boggy ponds, pools and marshes, often temporary and lacking fish; occasionally in slow streams.

Nature Notes:

Species name means "bearing forceps," perhaps alluding to the shape of the male's cerci.

Emerges earlier in season than the similar Northern Spreadwing and has much more restricted range in Rockies.

Distribution in Rockies somewhat uncertain because of long confusion of this species with Northern Spreadwing.

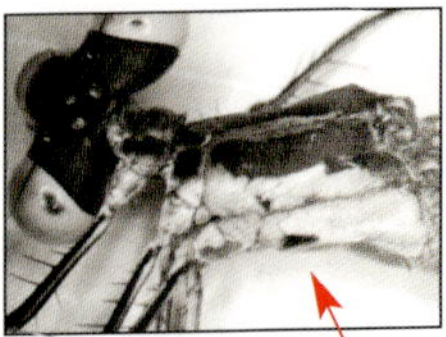

Look for the single spot under the thorax.

Description: Adults average 1.6 inches long.

A mid-sized, dark spreadwing. Thorax of male dark above with pale yellow or grayish-white sides. Lower side of thorax may have a single black spot (↑). Thorax has blue-green shoulder stripes. Thorax further darkens with age and is eventually obscured by pruinosity. Abdomen is dark with metallic green reflections and segments 9 and 10 are grayish-white pruinose. Wings clear. Female is similar, but her thorax sides are often light tan and her abdomen thicker. Female's ovipositor is long, extending to or beyond the tips of her cerci (↑).

Identification Clues: Separate male from male Northern Spreadwing by Sweetflag's two unequal "teeth" on inner margins of cerci (pg. 55); end tooth is smaller and blunter than basal tooth and teeth are further apart. See also Identification Clues on Northern page. Lyre-

The female Sweetflag Spreadwing is easy to spot because of her long ovipositor.

tipped male has S-shaped paraprocts (pg. 64) and has pruinosity on part of abdominal segment 8. Single black spot on lower side of thorax is helpful when present; Spotted Spreadwing has two spots there. Female's very long ovipositor is unmistakable, observation of tandem pairs therefore useful. Female Emerald Spreadwing's ovipositor is nearly as long, but is bright metallic green on top of her thorax and abdomen.

Distribution: Foothills and Montane Zone of extreme Northern Rockies and on east side of Continental Divide.

Life Cycle & Behavior: Adults emerge in June and July and fly into September. Eggs are laid into stems of Sweet Flag (genus *Acorus*), bulrushes and rushes above water level. Overwintering eggs can withstand air temperatures to about minus 4 degrees F. Snow cover is essential to protect eggs from very cold temperatures. Eggs hatch in spring and development of the nymph through about ten instars takes less than two months.

Northern Spreadwing *Lestes disjunctus*

APRIL	MAY	JUNE	JULY	AUG	SEPT	OCT

Breeding habitat: Mostly marshy or boggy ponds. Also bogs, fens and slow streams. Tolerant of acidic conditions and prefers fishless habitats.

Nature Notes:

An abundant and widespread *Lestes* throughout the Rocky Mountains at elevations that can exceed 11,000 feet.

Females occasionally lay eggs underwater, unique behavior among Rocky Mountain spreadwings.

Female lays more eggs per incision (up to six) than other pond spreadwings.

Two males may struggle while attempting to form a tandem pair with the same female.

Description: Adults average 1.5 inches long.

A mid-sized, dark, spreadwing. Thorax and abdomen of male is dark above with pale yellow or gray sides. Thorax has greenish-blue shoulder stripes (↑) and a thin pale line along top. Thorax darkens with age and gains some pruinosity. Abdomen is dark with metallic green reflections and segments 9 and 10 are grayish-white pruinose. Wings clear. Female is similar, with sides of thorax yellow to light tan and her abdomen is stout and black above. Rear of female's head is mostly dark with pale area only around neck. Female ovipositor is short, not extending beyond tip of abdomen.

Identification Clues: Two equal-sized "teeth" on inner margins of male cerci in dorsal view are diagnostic (pg. 55) Sweetflag Spreadwing male is similar, but has blunter end teeth. Sweetflag male is usually more pru-

Look at coloration on rear of head to distinguish females of Northern and Lyre-tipped Spreadwings.

inose on thorax than Northern and is much less common here. Lyre-tipped male has S-shaped paraprocts (pg. 64) and is pruinose on part of abdominal segment 8. Female Northern and Lyre-tipped are similar, but pale area on rear of head of Northern does not reach eyes; pale area usually reaches eyes on Lyre-tipped. See also distinctive stigma of Lyre-tipped. Sweetflag female has much longer ovipositor than Northern. Identifications of females may be uncertain.

Distribution: Widespread through all Rocky Mountain regions and zones except for Alpine.

Life Cycle & Behavior: Adults emerge in late June and July and fly into September. Males seize females in flight or when perched when they arrive at breeding sites. Eggs are laid in tandem, usually into the green stems of bulrushes above water level. Eggs are cold-tolerant, but need insulating snow cover; hatch in spring when water temperatures reach about 50 degrees F. Nymphs develop in 60 to 70 days and adults mature in 16 to 18 days after emerging.

Lyre-tipped Spreadwing *Lestes unguiculatus*

APRIL	MAY	JUNE	JULY	AUG	SEPT	OCT

Breeding habitat: Small, sunny ponds and marshes in open areas, usually temporary in some years. Avoids acidic habitats and those with fish.

Nature Notes:

Widespread throughout Rockies, but does not reach as high elevations as Spotted, Northern or Emerald Spreadwings.

Not an aggressive species; males may be chased by smaller damselfly species.

Species name means "small-clawed," perhaps referring to the clawed shape of the male's paraprocts. "Lyre-tipped" also refers to these distinctly shaped structures.

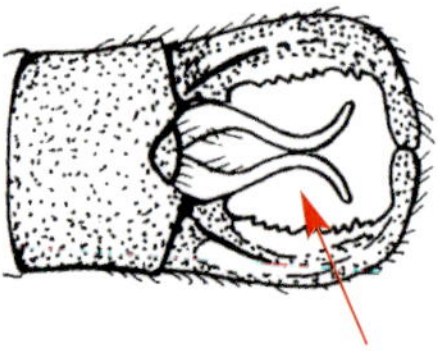

Description: Adults average 1.5 inches long.

Thorax of male is bronze above, pale yellow to whitish below and with pale green shoulder stripes. A thin pale line along thorax top may be evident. Thorax becomes increasingly pruinose with age. Abdomen metallic green, with grayish white pruinosity on segments 8, 9 and 10. Segment 8 partially pruinose, often showing a dark basal triangle (↑). Wings clear. Female's thorax is similar, but her abdomen is dark with greenish reflections on top and thicker. Tenerals of both sexes show more green than mature adults. Rear of female's head usually has extensive pale area that reaches eyes. Female's ovipositor is short, not extending beyond the tip of her abdomen.

Identification Clues: Male's S-shaped paraprocts (↑) with divergent tips are diagnostic in dorsal view. Note: slender, straight paraprocts

Look at coloration on rear of head to distinguish females of Lyre-tipped and Northern Spreadwings.

of Northern Spreadwing could be crossed causing an S-shaped look. Dark triangle on top of male's abdominal segment 8 is a fairly good field mark, but there is some variation. Both sexes have stigmas with pale ends. See also Identification Clues on Northern Spreadwing page. Identification of females may be uncertain.

Distribution: Foothills, Montane and lower Subalpine zones in all Rocky Mountain regions, but distribution limited in Southern Rockies.

Life Cycle & Behavior: Adults emerge from mid June through early July and fly into September. Males aggressively seize unpaired females (even dead females pinned to a stick). Eggs are laid in tandem into green emergent plant stems, often bulrushes or sedges. Choice stems are in small clusters or around the edges of larger beds—first to get knocked over and first to receive an insulating cover of snow. Overwintering eggs are cold-tolerant (to about minus 4 degrees F), but need snow cover. Eggs hatch in spring when water temperatures reach ~ 50 degrees F. Nymphs develop in about 60 days and adults mature 16 to 18 days after emerging.

Emerald Spreadwing *Lestes dryas*

APRIL	MAY	JUNE	JULY	AUG	SEPT	OCT

Breeding habitat: Small forest pools that dry up by summer (vernal pools); occasionally permanent ponds. Not found where aquatic predators are common.

Nature Notes:

Our earliest pond spreadwing to appear on the wing (usually by more than a week).

Thrives in shallow, temporary, woodland pools that lack fish and darner nymph predators.

A true Rocky Mountain warrior, widespread throughout all regions and sometimes found at elevations over 11,000 feet.

Species name refers to a mythical Greek wood nymph, perhaps indicating its preference for shaded, forest pools.

Ranges across Europe and Asia as well as North America.

Description: Adults average 1.5 inches long.

Our most striking pond spreadwing and the stockiest. Thorax and abdomen of male is bright metallic green above (↑) and pale yellow turning whitish pruinose below. Pale shoulder stripe thin or absent. Young males may appear more blue than green. Last two abdominal segments are grayish-white pruinose. Wings clear. Female is also metallic green, with thin, pale shoulder stripes and a thicker abdomen. She lacks pruinosity. Female's long ovipositor extends to or beyond tip of abdomen (↑) (like Sweetflag Spreadwing female).

Identification Clues: Boot-shaped paraprocts (widened tips) of male are diagnostic (pg. 55). Brighter green and heavier bodied than other spreadwings. Female could only be confused with female Sweetflag Spreadwing

The female shows the same bright green top of thorax and abdomen as the male, but her very long ovipositor extends beyond the tip of the abdomen.

which has an even longer ovipositor and the top of her thorax is dark, not bright green. Note however that tenerals of most spreadwing species may show metallic green in areas that are dark when mature.

Distribution: Montane and Subalpine zones.

Life Cycle: Earliest Rocky Mountain pond spreadwing emerges from early June through early July. Flight period usually ends by early August. Eggs are laid into green emergent plant stems, often bulrushes, sedges and grasses, above water. Eggs are large, hence the need for the female's large ovipositor. Overwinters as eggs in northern areas in late stages of embryonic development where they can withstand air temperatures lower than minus 4 degrees F. May overwinter as nymphs in milder climates. Eggs hatch in spring after pools fill with spring run-off. The nymphs develop extraordinarily quickly through ten instars in just a few months, so they can emerge before vernal pools dry up.

Pond Damsels
Family Coenagrionidae

The pond damsels comprise a huge family of mostly small, clear-winged species that delightfully display nearly all the colors of the rainbow. This family includes our smallest damselflies, some of which barely reach an inch long, and many of our most common species. Aspects of wing venation are the unifying features that set them apart from other families (see Family Key on pg. 44). Their stigmas are short, less than twice as long as wide. The male's terminal appendages are shorter than in other families, and the legs are quite short as well. Females often come in two color schemes: a male-like color pattern (homeochromatic) and a color pattern different than the male (heterochromatic).

Pond damsels inhabit the full spectrum of aquatic breeding habitats in the Rocky Mountains: lakes, ponds, marshes, swamps, bogs, fens, rivers, streams and spring seeps. Some species are habitat generalists that can be found at a wide range of aquatic habitats whereas others are closely tied to specific types of habitat (specialists). Adults can often be found near breeding sites around clearings, meadows, along roadways and railroad grades. As a group they fly from May through September in the Rocky Mountains, although most species have a short flight period lasting only four to six weeks.

Worldwide there are about 1,100 described species of pond damsels, with about 103 species in North America north of Mexico. This is easily the largest damselfly family in the Rocky Mountains, with 32 species in seven genera.

Dancers — Genus *Argia*

The Dancers are a colorful group of damselflies of medium build that range in size from fairly small to quite large. Males are predominantly blue or purple and females are various shades of brown, but male-like females are common in some species. Males may lose their bright colors when they get cold or when in tandem, but regain them when those conditions change. As with many odonates, colors of dancers will change as the adult ages, and the development of pruinosity may mask the underlying patterns. Dancers are known for their bouncy flight (hence their common name) and their tendency to alight on bare, sunny places such as roads, paths and shoreline rocks. Some species do like to perch on vegetation. Dancers are alert, tend not to rest long in one spot and may be challenging to approach closely. Most species engage in "wing-clapping," a behavior of quickly opening and closing their wings. Rocky Mountain species are diverse in appearance and are found in a variety of habitats but most often along flowing waters. A few species are locally common at warm or hot springs. Males arrive at breeding sites before females. They fly aggressively at other males, but without physical contact and with little clear defense of territories. Courtship behavior is also undeveloped in this group, with males seizing females upon their arrival at water. Females deposit eggs into wet wood, plant stems or in mats of floating plant material, usually at or just below the surface, often at communal sites. Females usually oviposit in tandem, with males usually guarding in the "sentinel" position, but females may be unattended. Powdered Dancers oviposit well underwater, the male sometimes remaining in tandem and becoming completely submerged. Nymphs live in gently moving water either under rocks in rapids or in mud and plant material in quieter areas. The nymphs are easy to recognize with their short, stocky bodies and wide caudal lamellae. Beyond that, the nymph stage is poorly known for most species. The most likely meaning for the name *Argia* is "lazy," but no one seems to known how this applies to the group. *Argia* may have been intended to mean "bright," from the Greek *argos*, referring to the coloration of the males.

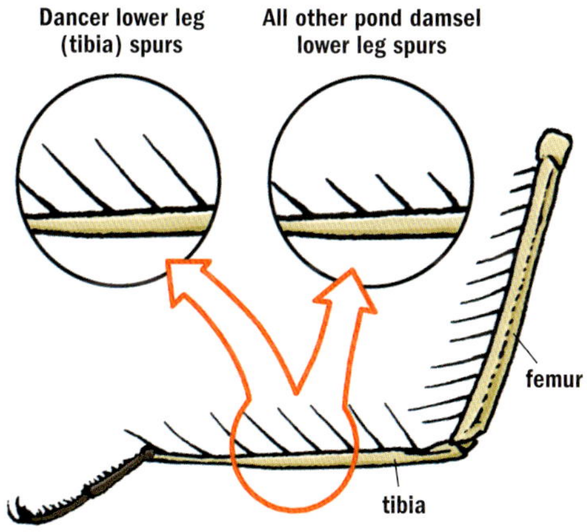

Easy dancer identification: leg spurs are twice as long as the space between them. All other pond damsels have spurs about as long as the space between them.

Although some dancers are similar in size and coloration to Eurasian and American bluets, they are easily separated from these groups by a number of behaviors and physical characteristics. One distinctive habit of dancers is to perch with wings held high, slightly above the abdomen, and not along side touching it, as do bluets. Dennis Paulson has pointed out another remarkable perching difference between the groups: when perched on vegetation overhanging water, dancers tend to face the stream whereas bluets tend to face the shore. Among the physical differences between the groups are the long spurs on the lower legs (tibiae) of dancers that are twice as long as the spaces between them. Bluets have tibial spurs only about as long as the spaces between them. The dark shoulder stripes of dancers are either notched or forked near the upper ends, or are distinctly narrowed near the middle. In contrast, the dark shoulder stripes of bluets are never forked and are either the same width throughout or are gradually narrowed toward the upper end. The terminal appendages of male dancers have cerci shorter than their paraprocts; the reverse is often true of bluets. Dark markings of the abdomens of females are often on the sides of segments of dancers versus on the top of segments of bluets. Female dancers lack a vulvar spine on abdominal segment 8; American bluet females (genus *Enallagma*) have noticeable vulvar spines.

Males of our species are easily identified in the field, or in the hand, by their colors, patterns and the shapes of their terminal appendages. Females can usually be identified in the hand. This is a Neotropical genus of about 111 species, most living in Central and South America. Currently, 32 species are recognized in North America, with nine occurring in the Rocky Mountains. By the way, when I tell my wife I'm going out to watch dancers, she knows I mean the winged variety!

Eurasian Bluets — Genus *Coenagrion*

The Eurasian bluets are a northern group of small to medium-sized damselflies that I find to be exceptionally beautiful. Males of the Taiga Bluet are especially compelling, with shades of blue and green on their thorax and abdomen that are hard to adequately describe. Males of most species are primarily blue and black and can easily be mistaken for male American bluets (genus *Enallagma*) or even forktails (genus *Ischnura*). Females are marked similar to the males and come in two basic color patterns: male-like hues (homeochromatic) and duller, greenish or brownish hues (heterochromatic). Adults tend to fly low and perch often among vegetation. Rocky Mountain species are usually found along the edges of ponds, marshes and bogs. Males arrive at breeding sites before females. They are rarely aggressive with other damselflies and do not defend territories. Females oviposit in tandem on emergent or floating vegetation with males alertly guarding in the "sentinel" position.

Nymphs live in still, often shallow waters where they crawl about on submerged vegetation hunting their prey. The nymphs are small and brown or greenish and are very similar to the nymphs of American bluets, from which they are difficult to separate. North American species generally complete their life cycle in one year, but in the Far North development is slower, possibly taking as long as three or four years. This genus is better able to tolerate colder climates than any other group. A fascinating study found that the nymphs of the Taiga Bluet overwintered embedded in five to eight inches of ice! They were often found completely encased in an upside-down position as though trapped while walking about on the undersurface of the ice, and when the ice melted they resumed activity. The ability to overwinter in this way allows them to survive in shallow northern ponds that freeze to the bottom in winter and protects them from predators.

This is a circumpolar boreal group, with most species occurring in Europe and Asia. Only three species are found in North America, two of which are found in the Rockies. The Taiga Bluet is widespread and common throughout the Rockies in all regions and at all elevations except for the Alpine Zone. The Subarctic Bluet is one of our rarest damselflies, occurring only in the northern tip of the Northern Rockies. Eurasian bluet males differ from male American bluets in the shapes of their terminal appendages. Females can be separated in the hand from American bluets by their lack of a vulvar spine on abdominal segment 8 (all American bluet females have it). Our males can usually be identified to species in the field by their colors and patterns. The shapes of their terminal appendages, featuring paraprocts that are bilobed in side view, are diagnostic in the hand. Females can be trickier to separate but can usually be identified in the hand. Black markings on the top of the abdominal segments of females are usually bluntly rounded or squared off, not pointed like bullets toward the thorax as on many female American bluets.

American Bluets — Genus *Enallagma*

The American bluets are a large genus of small to mid-sized damselflies that are abundant throughout the Rockies around a wide variety of mostly still-water habitats. They are commonly called "bluets" because the males of most species are bright blue with contrasting black markings that consist of distinct stripes on the thorax and bands on the abdomen. While glancing through the field notebook of an Ontario friend many years ago I saw reference to "lots of BBBs" and asked him what that stood for. "Black and Blue Boys," he replied with a smile, referring to the males of this genus. Unlike many damselflies, males remain colorful throughout life, as there is little tendency for the brilliant blues to darken or become obscured with pruinosity as they age. Males can sometimes be identified in the field by their color patterns, but firm identifications often require examination in the hand, where the distinctive shapes of their terminal appendages can be seen in side view with a hand lens (pgs. 76-77). Even when color patterns seem sufficient to identify a species, it is always a good idea to confirm it by comparing the side view of the cerci and paraprocts with the drawings in this guide. If you get into the habit of doing this early, you won't regret it!

Female American bluets are usually duller than males in color, typically showing olive, yellowish green or brown hues where the males are blue. However, females of many species have male-like blue forms, which are usually less common than the typical heterochromatic form. Females are often blacker on top of the abdominal segments than the males, with many species having torpedo- or bullet-shaped markings on most segments that point toward the thorax. Females can sometimes be identified in the hand, or even in the field, by distinctive color patterns, but some "clusters" of species are very similar, requiring careful examination of their **mesostigmal plates** (a.k.a shoulder pads) under a microscope. Females of these species can often be identified (tentatively) by the males they are associated with. Be aware though, that males may take in tandem females of other species, so look for mating pairs! Females have a vulvar spine, contrasting with the dancers and Eurasian bluets, which lack the spine. Wings of both sexes are clear, moderately "stalked," and have dark stigmas, unless noted differently in the species accounts. Eyespots (postocular spots) on top of the head are always present and are sometimes helpful in identification because they vary in size among species. The tibial spurs are short, in contrast with the dancers.

The pattern of dark and pale thoracic stripes can help to identify some American bluets.

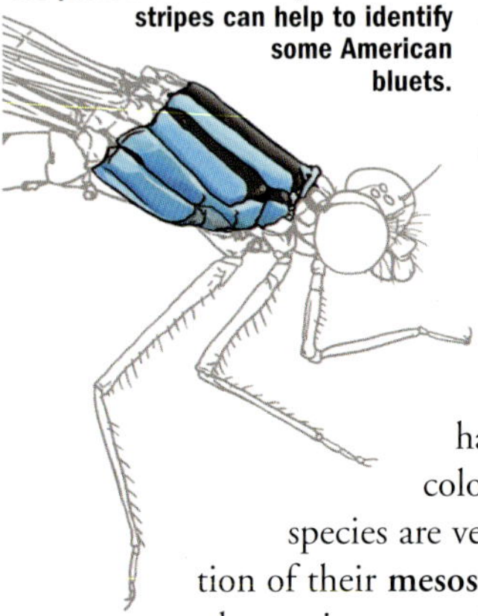

These familiar, bright-blue damselflies often abound in grasses and sedges around ponds and lakes throughout the summer. I suspect John Acorn had this genus in mind when he gave his Alberta damselfly guide the delightful subtitle "Flying neon toothpicks in the grass." Well-vegetated still-waters are the habitats of choice for most species, but other habitats are sometimes used, including streams and rivers. Most American bluets are weak fliers, making short forays close to vegetation after tiny insect prey, although some do fly strongly over water. Males do not usually defend territories or show much aggression toward other males. Females usually oviposit while in tandem but sometimes alone, into living or rotting plant stems or algal mats, sometimes well underwater.

The nymphs are green or brown and of moderate build, with gills that are variable in shape, though usually narrow. Some species do best in waters lacking fish. All Rocky Mountain species overwinter as nymphs, but the nymph stage of some species is not well known. This worldwide genus has more species in North America (at least 37 of them) than are found anywhere else. A diverse group of 10 species inhabits the Rocky Mountains.

Forktails — Genus *Ischnura*

Forktails are small, strikingly patterned damselflies that are similar in appearance to the American bluets (genus *Enallagma*) and the Eurasian bluets (genus *Coenagrion*). However, their reduced wing venation, the variation in color of females and the small forked structure at the tip of the abdomen of males set the genus apart from other closely related groups. Males of all species (except Citrine Forktail) have dark abdomen with blue tips, which along with a green-striped thorax is a distinctive feature of forktails. Male forktails are also unique in often having shape and color differences of stigmas between fore and hind wings. But the signature feature of male forktails is exactly that—the forked projection on top of abdominal segment 10 that the famous odonatist E. M. Walker aptly called a "turret-like eminence, which is more or less distinctly bi-lobed." This forked projection varies greatly in size among species. Females in this genus come in a bewildering array of colors for two reasons. First, all of our species except for the Citrine Forktail have a male-like form (homeochromatic) in addition to the typical female (heterochromatic) form. Then, to add to the confusion, heterochromatic females change color as they age. Most are orange or orange-brown when young, but darken or become covered with a grayish-white or bluish pruinescence as they age, which may be associated with reduced sexual receptivity. This pruinescence gradually obscures much, or all, of the underlying color pattern. The scientific name Ischnura means "slender-tailed," referring to their slender abdomen.

In addition to their distinctive appearance, forktails also have some unusual life history features that set them apart from other groups. Our species have long flight periods; emerging in the spring and flying throughout the summer. Spring emerging specimens tend to be larger, with smaller individuals appearing in late summer and autumn. Newly emerged forktails reach reproductive maturity more rapidly than other damselflies, often in a few days and sometimes in just one day. Females of some species may mate only once (unusual for damselflies) and females of most of our species typically oviposit alone (Black-fronted is our only exception) into plant material at or just below the water surface. A European species appears to hold the world record for longest copulations—up to 7.5 hours! Long copulations are a strategy males use to monopolize females until too late in the day for them to mate with other males. Forktails are feeble fliers that hold close to the emergent vegetation along the shoreline of their breeding habitat.

This abundant genus is found throughout the world in still-water habitats. There are 14 species in North America, of which eight are residents of the Rocky Mountains. The Pacific, Plains and Western forktails are widespread and often abundant throughout the Rockies. The remaining species are either uncommon here or are regionally distributed. In the field, males are readily placed into groups of similar species by their distinctive colors and patterns. Among these groups, Eastern, Western and Mexican Forktails look very similar to each other, as do Pacific and Plains Forktails. Species within these groups should be netted so identifications can be confirmed by viewing the terminal appendages in side view with a hand lens. Males of Desert, Black-fronted and Citrine forktails all have a distinctive appearance. Identifying females is usually not easy. Sometimes they can be identified in the field by their distinctive color patterns, but the pruinescence on older females largely obscures these patterns. Identification of problematic females can only be done with certainty by carefully examining their mesostigmal plates under a microscope. Tentative identifications of females can often be made by noting the males with which they associate.

Painted Damsel — Genus *Hesperagrion*

This genus has just one species in the whole wide world, and boy is it a dandy! We are fortunate to have it in scattered areas of the Southern Rockies. Male is unlike any other Rocky Mountain damselfly both in its spectacular coloration and shape of its cerci (upper clasper). Behavior is generally forktail-like, but body structures not so.

Red Damsels — Genus *Amphiagrion*

This beautiful little genus contains only two species in North America: Eastern Red Damsels and Western Red Damsels. Interestingly, populations in the Midwest are intermediate between Eastern and Western Red Damsels and debate continues about how many species should be named. Specimens throughout the Rockies, including on the east slope of the divide, are consistent with the western species, and I will refer to them as such until any new study results come forth. Nymphs of this genus are distinctive in our region in having sharply angled rear corners of the head. Genus name means "both agrion (a word for damselfly)," probably referring to the two species in the genus.

Sprites — Genus *Nehalennia*

The sprites are tiny, dainty damselflies that are easy to identify once you find their haunts. Their small size, unstriped metallic-green thorax and lack of eyespots sets them apart at a glance from all other damselflies. Sprites typically fly low among plants and are reluctant to fly over open water. For this reason and because they are so small, they are inconspicuous and require a bit of patience to observe. Once you find them, they are fun to watch and will often allow you to approach quite closely. These fragile insects are both beautiful and interesting. The name *Nehalennia*, appropriately enough, is after a river goddess of the Rhein. I've noticed that when shown their first sprite, people often respond by saying, "How cute!" or "Isn't it adorable!" Six species of sprites are known from around the world, five from the Americas and just one species is found in the Rocky Mountains.

Abdomen tip side views of male Pond Damsels

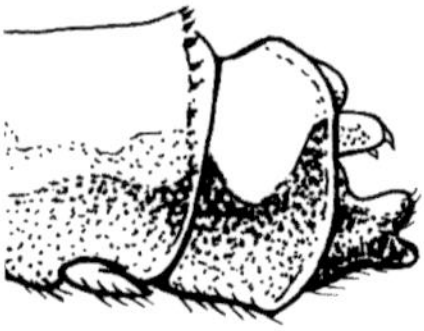

Blue-fronted Dancer
Argia apicalis pg 78

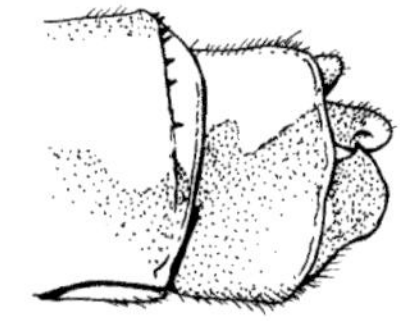

Powdered Dancer
Argia moesta pg 80

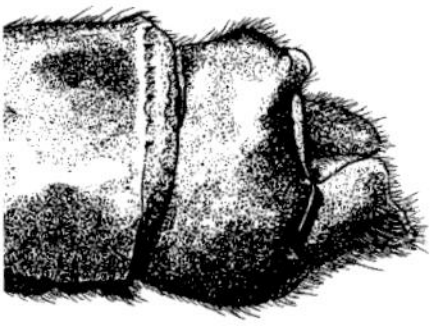

Blue-ringed Dancer
Argia sedula pg 82

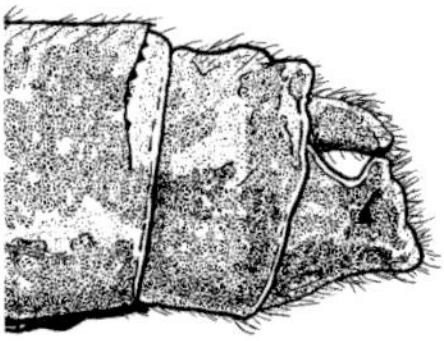

Paiute Dancer
Argia alberta pg 84

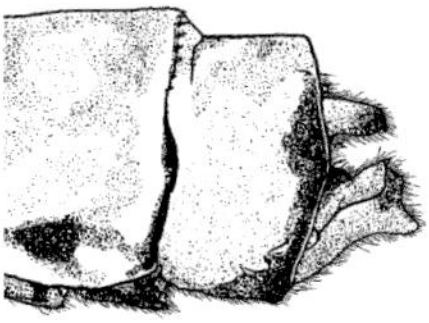

Emma's Dancer
Argia emma pg 86

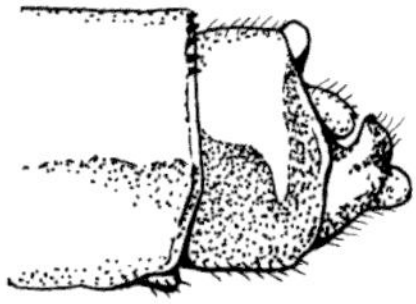

Violet Dancer
Argia fumipennis pg 88

Springwater Dancer
Argia plana pg 90

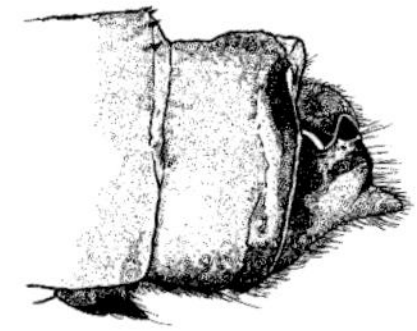

Vivid Dancer
Argia vivida pg 92

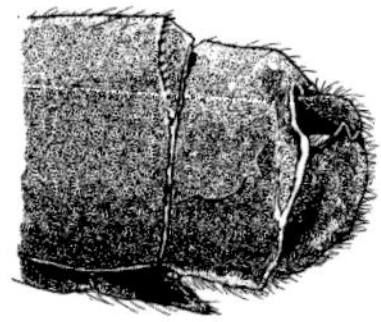

Sooty Dancer
Argia lugens pg 94

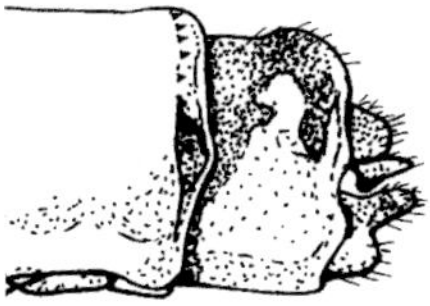

Subarctic Bluet
Coenagrion interrogatum pg 96

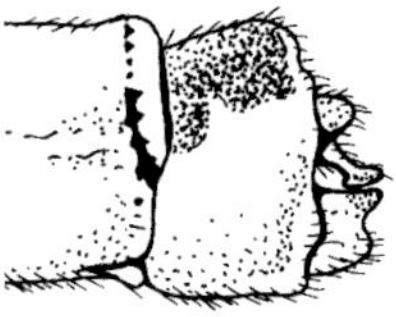

Taiga Bluet
Coenagrion resolutum pg 98

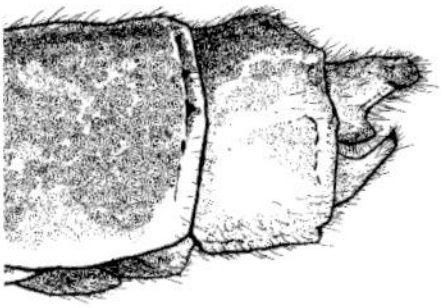

Arroyo Bluet
Enallagma praevarum pg 100

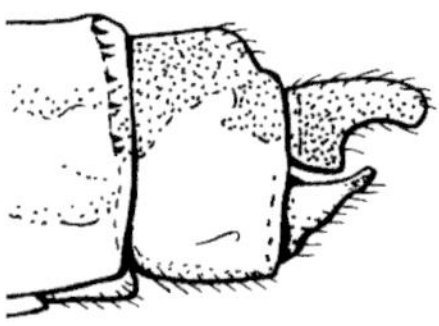

River Bluet
Enallagma anna pg 102

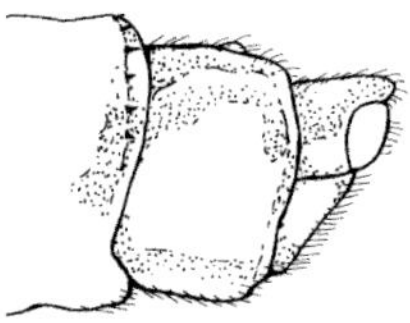

Familiar Bluet
Enallagma civile pg 104

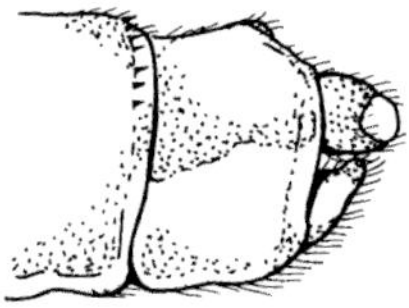

Tule Bluet
Enallagma carunculatum pg 106

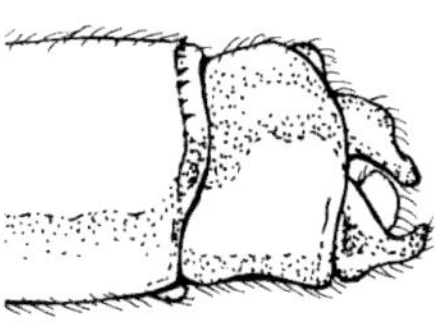

Alkali Bluet
Enallagma clausum pg 108

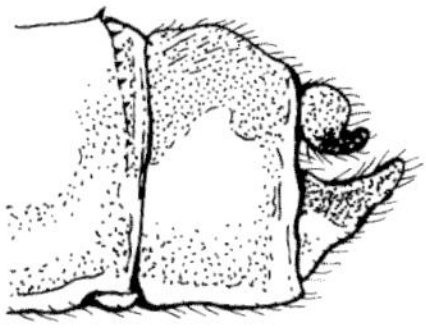

Northern Bluet
Enallagma annexum pg 110

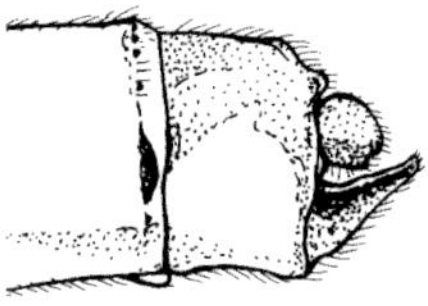

Boreal Bluet-
Enallagma boreale pg 112

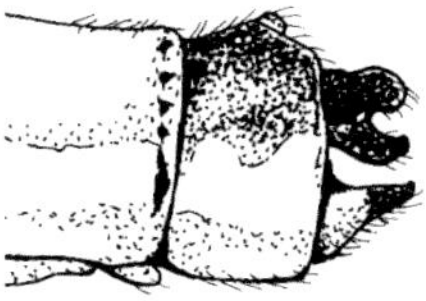

Marsh Bluet
Enallagma ebrium pg 114

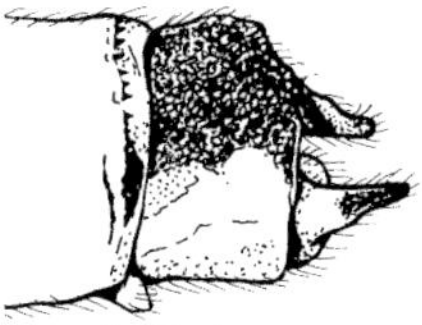

Hagen's Bluet
Enallagma hageni pg 116

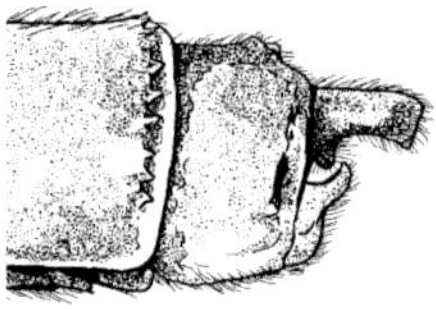

Double-striped Bluet
Enallagma basidens pg 118

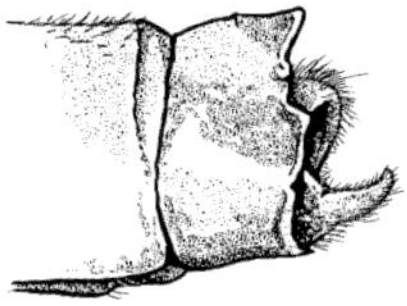

Desert Forktail
Ischnura barberi pg 120

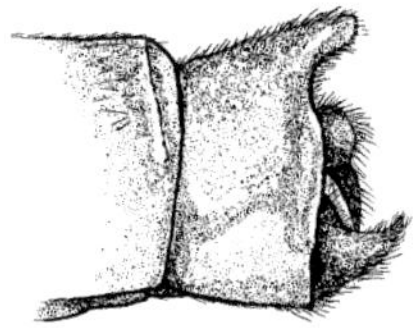

Pacific Forktail
Ischnura cervula pg 122

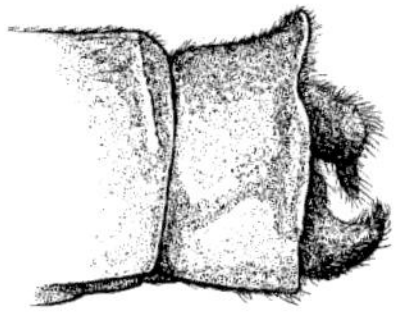

Plains Forktail
Ischnura damula pg 124

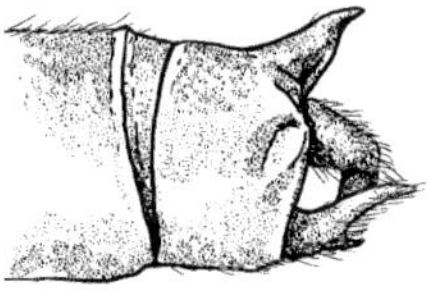

Mexican Forktail
Ischnura demorsa pg 126

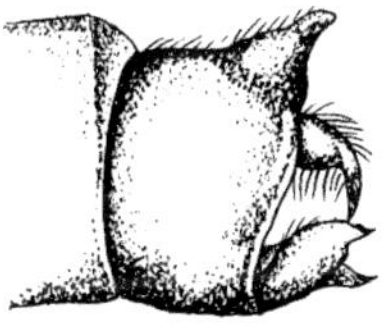

Western Forktail
Ischnura perparva pg 128

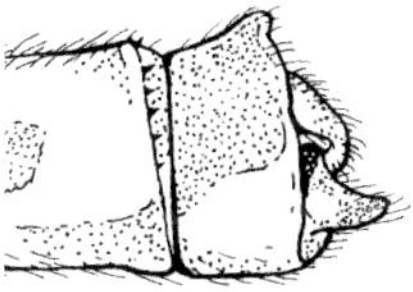

Eastern Forktail
Ischnura verticalis pg 130

Black-fronted Forktail
Ischnura denticollis pg 132

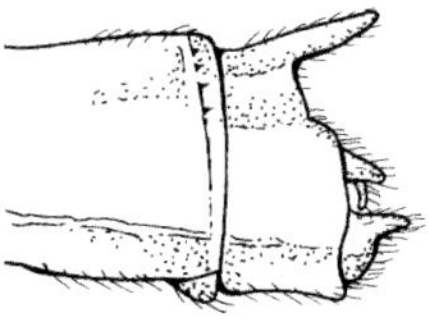

Citrine Forktail
Ischnura hastata pg 134

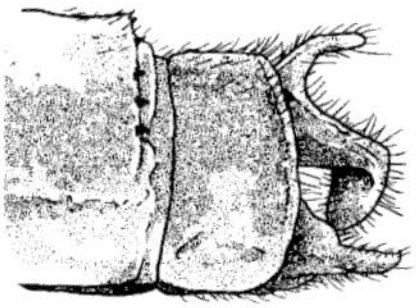

Painted Damsel
Hesperagrion heterodoxum pg 136

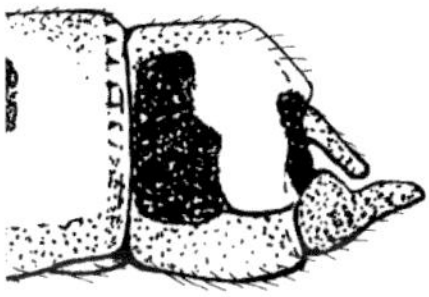

Western Red Damsel
Amphiagrion abbreviatum pg 138

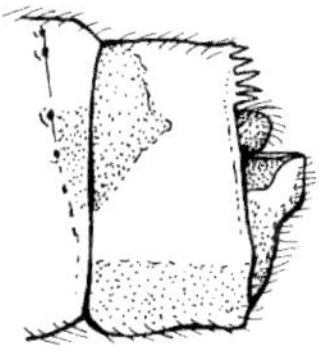

Sedge Sprite
Nehalennia irene pg 140

Blue-fronted Dancer *Argia apicalis*

APRIL	MAY	JUNE	JULY	AUG	SEPT	OCT

Breeding habitat: Mid-sized to large rivers with gentle current and sand or mud bottoms; occasionally streams, lakes and ponds.

Nature Notes:

Flies with a "dancing" movement over streamside vegetation.

Male's bright blue thorax is easily seen when he is perched on bare ground.

Males and females in tandem may "explore" for a good egg-laying site for nearly an hour after mating.

Male's terminal appendages: cercus has tiny down-pointing teeth.

Description: Adults average 1.5 inches long.

A large, brightly colored and slender species. Males show bright blue on the thorax (↑) and on top of last three abdominal segments while the rest of abdomen is dark. Dark shoulder stripes are hairline thin. A small black blotch is usually visible on lower thorax above middle leg (smaller or wanting on female). His eyes and face blue. Female is either brown on thorax or blue like the male, with dark shoulder stripes hairline thin or absent. Her abdomen is dark on top with a narrow pale stripe on some segments and dark side stripes on segment 9. Wings of both sexes clear.

Identification Clues: Bright blue thorax of male with only a hairline shoulder stripe is unmistakable. Shape of the male cercus is also distinctive (↑). Brown-form female is similar to Powdered Dancer and Emma's Dancer, all of which have very thin shoulder stripes. Female Blue-fronted has only one cell in wing directly below the stigma whereas female

Blue-form females like this one bear some resemblance to males.

Powdered usually has two cells there (with a crossvein below middle of stigma) and Blue-fronted has more heavily striped abdomen tip. Female Emma's has abdomen not as heavily marked with dark stripes on sides of segments as Blue-fronted.

Distribution: Foothills and lower Montane Zone on eastern slope of Southern Rockies.

Life Cycle & Behavior: Adults have a long flight period, beginning to emerge in mid June and flying into September. Emergence occurs around midday, just a few inches from the water's edge. Individual adults live about a month at most. Often perches on bare ground, horizontally when on vegetation. Males are mildly territorial and may interact aggressively with one another, but only to space themselves along a shoreline, not to defend specific egg-laying sites. Females often congregate to oviposit below the water's surface into well-decayed logs. Overwinters as a nymph in one of the last instars.

Powdered Dancer *Argia moesta*

APRIL	MAY	JUNE	JULY	AUG	SEPT	OCT

Breeding habitat: Rock-margined streams, rivers and lakes.

Nature Notes:

Only whitish damselfly in Rockies due to the male's extensive pruinosity.

Many pairs may oviposit together in dense breeding aggregations.

Male remains in tandem (at least part of the time) when completely submerged while female lays eggs; unusual for a damselfly.

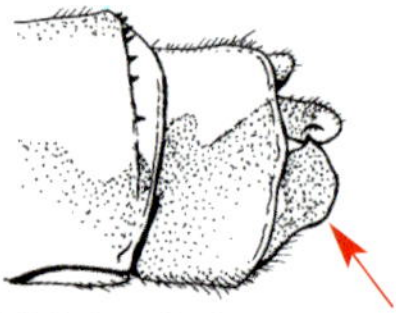

Male's terminal appendages: paraprocts unbranched in side view.

Description: Adults average 1.6 inches long.

A large light-colored dancer. Occurs in a wide array of colors and markings depending on age and gender. Male becomes almost completely whitish with pruinosity (↑) when mature, except for mid-abdominal segments, which remain dark. Immature males are tan to dark brown, turning darker with age before becoming pruinose. Wide, dark shoulder stripes of male become obscured by pruinosity. Females are blue-form or brown-form, with hairline dark shoulder stripes on thorax (↑). Her abdomen with pale, wide top stripe and narrow dark side stripes, often noticeably pale at tip (↑). Wings of both sexes clear.

Identification Clues: Extensive whitish pruinosity of male is unique. Sooty Dancer also becomes largely pruinose but is dark. Our Powdered and Sooty only dancers with two

The blue-form female shows the characteristic pale abdomen tip.

cells below the stigma of the wing, separated in the middle by a crossvein (both sexes). Paraprocts in side view not branched (↑) (Sooty also this way). Female very similar to female Blue-fronted Dancer, but blue-fronted lacks crossvein below middle of stigma and abdomen tip is more heavily striped.

Distribution: Foothills and lower Montane Zone in southeastern tip of Central Rockies and west slope of Southern Rockies.

Life Cycle & Behavior: Adults have a long flight period, beginning to emerge in early June and flying into September. Tenerals mature two weeks after emerging; adult life span only about a month. Often perches on bare ground. More likely than other dancers to hold wings low, along side abdomen. Pair descends, usually in tandem, sometimes deep underwater to oviposit in wet wood, plant stems or algae-covered rocks and may remain submerged for an hour or more (male often disengages after 5 to 20 minutes). Nymphs cling to stones and other bottom debris, sometimes in fast current. Overwinters as a nymph, likely in one of the last instars.

Blue-ringed Dancer *Argia sedula*

APRIL	MAY	JUNE	JULY	AUG	SEPT	OCT

Breeding habitat: Wide variety of small to large streams and rivers; occasionally lake shores and ditches.

Nature Notes:

More likely to perch on streamside vegetation, facing stream, than on ground or rocks.

Often perches higher on vegetation than other dancers.

Male's terminal appendages: cerci are long and paraprocts show shallow branch in side view.

Description: Adults average 1.3 inches long.

A small, attractive dancer that could easily be mistaken for a bluet. Males show dark blue on thorax with heavy black top and shoulder stripes. Thick black shoulder stripe slightly notched at upper end (↑). Blue on thorax becomes lighter on lower sides. His abdomen mostly black with pale blue rings at bases of most segments. Top and sides of last three abdominal segments blue with black streaks along extreme lower sides. Male's eyes and face dark blue. Female completely unlike male, is light brown with olive highlights on thorax and abdomen. Her dark top and shoulder stripes very thin. Her abdominal segments with slightly deeper brown apical rings, fading toward lightened tip. Wings of male clear or with very light amber tint. Wings of female with noticeable light brown tint.

The female is quite pale in coloration.

Identification Clues: Males most likely to be confused with some male American bluets, but Rocky Mountain bluets have thinner black shoulder stripes not notched at upper end. Male's mostly black abdominal segment 2 is distinctive. Paiute Dancer is similar but Blue-ringed has narrower pale shoulder stripes and different pattern of blue rings on abdomen (compare photos). Paiute has all black top of abdominal segment 7; Blue-ringed has blue ring at base of that segment. Male terminal appendages viewed in hand also distinctive (↑). Female is less patterned than most brown female damselflies in Rockies, especially at abdomen tip. Amber-tinted wings also help distinguish females.

Distribution: Foothills and lower Montane Zone in northern and extreme southern areas of Southern Rockies.

Life Cycle & Behavior: Adults emerge primarily in June and may fly into September. Eggs are laid in tandem into green plant stems at or near waters' surface or into floating vegetation. Overwinters as a nymph in one of the last several instars.

Paiute Dancer *Argia alberta*

APRIL	MAY	JUNE	JULY	AUG	SEPT	OCT

Breeding habitat: Small sedge-bordered streams and marshy springs, often around hot springs.

Nature Notes:

Common name after the Paiute Indians of the Great Basin.

A denizen of warm and hot springs in northern areas of its Rocky Mountain range, the Paiute Dancer is more likely to be found along sandy, sedge-lined creeks in the Southern Rockies.

Scientifically named by Clarence Kennedy after his father Albert.

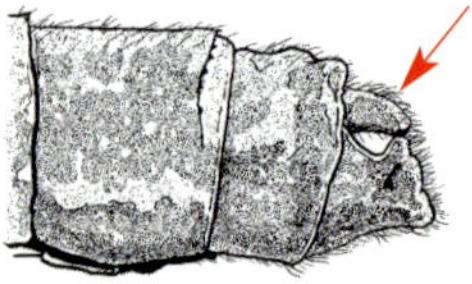

Male's cerci are elongated

Description: Adults average 1.2 inches long.

A small, handsome, blue and black dancer. Adult males show much blue on thorax with fairly heavy black top and shoulder stripes. Black shoulder stripe deeply forked, fork splitting about middle of stripe and incomplete at upper end (↑). His abdomen mostly black with pale blue rings at bases of most segments and apices of some. Top and sides of last three abdominal segments blue, black very low on sides. Male's eyes and face dark blue. Female's thorax either light brown/gray or blue-form like male. Her dark shoulder stripes thinner than male, and often noticeably forked starting about middle. Her abdominal segments light brown with dark streaks on sides of segment and deeper brown apical rings. Wings of both sexes clear.

Identification Clues: Males most likely to be confused with some male American bluets,

Most females have a forked, dark shoulder stripe, but this female lacks the fork—rarely the case—making her tough to identify.

but bluets have black shoulder stripes not forked. Blue-ringed Dancer similar but Paiute has wider blue shoulder stripes, less black on abdominal segment 2 and all black on top of segment 7 (Blue-ringed has blue ring at base of that segment). The small size, dark coloration of the abdomen and straight, elongate cerci (↑) separate the males from all other Rocky Mountain dancers. Female differs from other brown or blue-form Rocky Mountain dancers (except Violet) in having dark shoulder stripe forked and black streaks on sides of all abdominal segments. Black streaks on sides of segments shorter on Paiute than on Violet (compare photos).

Distribution: All Rocky Mountain regions in Foothills, Montane and lower Subalpine zones.

Life Cycle & Behavior: Adults emerge primarily in June and July and may fly into October. Adults not usually seen in large numbers. Eggs are laid while pair is in tandem. Life history has been scarcely studied.

Emma's Dancer *Argia emma*

APRIL	MAY	JUNE	JULY	AUG	SEPT	OCT

Breeding habitat: Clear, medium-sized or larger streams and rivers, often rapid and rocky with open canopy; sometimes lakes.

Nature Notes:

Tandem pairs often stay engaged for hours on a sunny shore.

May rest on railroad tracks in the sun.

Primarily a species of rapid mountain streams.

Along with Vivid Dancer, one of the most common and widespread Rocky Mountain dancers.

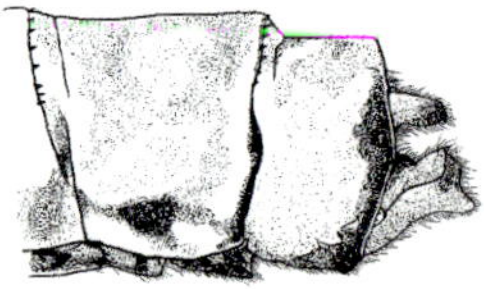

Male's terminal appendages.

Description: Adults average 1.5 inches long.

A large, handsome dancer. Thorax and abdomen of male is mostly violet, with black shoulder stripes unforked and hairline thin in middle (↑). Abdominal segments have black apical rings increasing in size toward the abdomen tip. Last three segments powder blue on top with some black low on sides. Eyes and face purple. Female either light brown all over or pale olive on thorax and abdomen base. Her thorax with very thin top and shoulder stripes. Abdominal dark markings like male, but tip is unmarked tan. Females are pale for dancers, with few dark markings. Wings of both sexes clear.

Identification Clues: Male is easily identified in the hand by shape of cerci in dorsal view (curves outward). In the field, combination of purple color, thin top thorax stripe,

This female has beautiful though subtle aqua tones along her mid-abdominal segments.

unforked shoulder stripe, blue abdomen tip and relatively large size set it apart from similar Rocky Mountain dancers (Violet Dancer and Purple-form Springwater Dancer). Female is least marked of all our dancers, but positive identification may require microscopic look at mesostigmal plates.

Distribution: Widespread throughout most regions in Foothills and Montane Zone, but uncommon in Southern Rockies.

Life Cycle & Behavior: Adults begin emergence in June and continue through July; may fly into September. Often abundant at prime habitats. Eggs are laid in tandem into a wide variety of aquatic vegetation, including willow roots, at or not far below the water surface, often at communal oviposition sites. Male may remain in tandem and become submerged while female oviposits. Males contact guard in sentinel position. Perches on bare ground or rocks in sunny areas more often than on vegetation. Overwinters as a nymph, likely in one of the last instars. Nymphs crawl barely out of water to emerge.

Violet Dancer *Argia fumipennis violacea*

APRIL	MAY	JUNE	JULY	AUG	SEPT	OCT

Breeding habitat: Clear, gentle, streams and stream backwaters; also some ponds and lakes.

Nature Notes:

Male is strikingly attractive—one of only three mostly purple damselflies in the Rockies.

Male assumes "sentinel" position while female oviposits (see opposite page photo).

Flies most actively in morning and during midday.

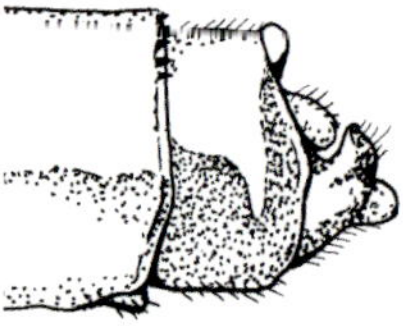

Male's terminal appendages.

Description: Adults average 1.2 inches long.

A small, easily identified, purple dancer. Thorax and abdomen of male is mostly violet, with a black, forked shoulder stripe (↑) on each side and narrow black rings on most abdominal segments. Younger male is violet-blue. Tops of segments 8, 9 and 10 are violet blue to powder blue. Female is mostly brown with thin, forked shoulder stripes and longitudinal black stripes on most abdominal segments (↑). Wings of both sexes clear in Rockies. Two subspecies that occur further south and east have smoky or very dark wings.

Identification Clues: Violet blue to powder blue abdomen tip, forked shoulder stripes and small size of male easily separate it from other purple dancers. Female separable from most other dancers by small size, thin, forked shoulder stripes and abdomen tip light-colored on

Note the thinly-forked dark shoulder stripe on this female.

Male in "sentinel" position while female oviposits.

top with dark stripes on side. Could be confused with female Paiute Dancer, but usually has longer dark stripes on sides of abdomen than Paiute. Female may not always be identifiable—look at male when mating.

Distribution: Rare and sparsely distributed in Foothills and lower Montane Zone of Central and Southern Rockies, mostly on east slope of Continental Divide.

Life Cycle & Behavior: Adults have a long flight period, beginning to emerge in mid June and flying late into September. Perches on ground, rocks or vegetation. Pair oviposits in tandem on wet logs and other plant materials and rarely descends below water's surface. Nymphs lurk on plant stems and on other bottom debris. The nymph stage is not well known. Overwinters as a nymph, probably in one of the last instars.

Springwater Dancer *Argia plana*

APRIL	MAY	JUNE	JULY	AUG	SEPT	OCT

Breeding habitat: Shallow streams, mid-sized and smaller, including spring and hillside seeps.

Nature Notes:

Once considered a subspecies of Vivid Dancer, hence their great similarity.

Males holding best territories have greater chances than other males of seizing females as they fly toward breeding sites.

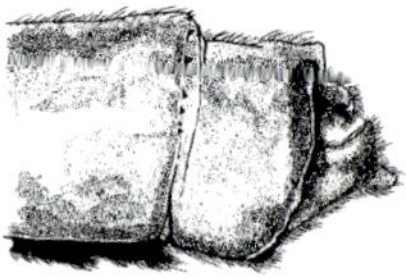

Male's terminal appendages.

Description: Adults average 1.4 inches long.

Males of this jaunty dancer occur in two color forms: blue-form is the norm in most of its range; purple-form occurs in Arizona and parts of New Mexico. Rocky Mountain specimens are mostly purple-form but could be either. Thorax and abdomen of male is mostly purple (blue in blue-form), with a black shoulder stripe unforked (rarely forked) and hairline thin in middle. Top thoracic dark stripe of moderate width. Middle abdominal segments have black apical rings increasing in size away from body and may have black streaks of variable size on sides. Segment 7 mostly black. Last three segments light purple or bluish. Base color of female can be brown, purple or blue depending on population. Her markings on thorax similar to male. Her abdomen with black blotches at apices of middle segments

Note the bright blue abdomen tip of this brown-form female.

and black streaks along the sides. Segment 7 mostly dark. Segments 8-10 entirely pale blue. Wings of both sexes clear.

Identification Clues: On purple-form male, look for thin, unforked shoulder stripe and abdomen tip that may show some purple (can be bluish). Blue-form male extremely similar to Vivid Dancer and is consistently separable from Vivid only by subtle differences of the cerci seen under a microscope. Females cannot be reliably distinguished from female Vivid except under a microscope. See Vivid Dancer for other differences.

Distribution: Rare in Foothills and lower Montane Zone of southern third of Southern Rockies.

Life Cycle & Behavior: Adults emerge primarily in June and may fly into September. Males are mildly territorial which serves to space them along shorelines. Female lays eggs in tandem into a wide variety of plant material at or near water surface. Nymphs lurk on plant stems and on other bottom debris. The nymph stage is not well known. Overwinters as a nymph, probably in one of the last instars.

Vivid Dancer *Argia vivida*

APRIL	MAY	JUNE	JULY	AUG	SEPT	OCT

Breeding habitat: Shallow, well-vegetated streams, mid-sized to much smaller, including spring runs; also hot springs.

Nature Notes:

"Vivid" is a very apt description of its color; first time I saw one I couldn't believe how intense the blue was.

Regularly associated with springs and spring-fed streams.

Along with Emma's Dancer, one of the most common and widespread Rocky Mountain dancers.

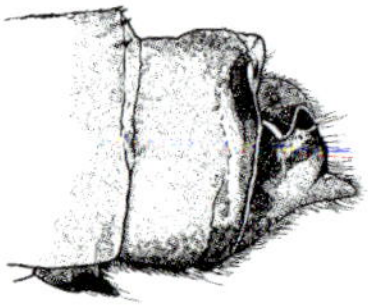

Male's terminal appendages.

Description: Adults average 1.4 inches long.

Thorax and abdomen of male is mostly vivid blue, with black top thorax stripe and black shoulder stripes usually unforked and hairline thin in middle. Middle abdominal segments have black apical rings increasing slightly in size away from body and black streaks of variable size on sides. Segment 7 mostly black above. Last three segments entirely blue. Eyes and face blue. Female either light brown all over or blue on thorax and abdomen tip like male. The two color forms about equally common. Her thorax striped like male. Her abdominal dark markings more extensive than male on sides of segments, but tip is unmarked pale (tan or blue). Segment 9 may have small black spots at bases (not always present). Wings of both sexes clear.

Identification Clues: Male is the only large blue dancer throughout most of the Northern and Central Rockies. Vivid is hard to distin-

This blue-form female was photographed on a rock—a very characteristic perch for this species.

guish from blue-form Springwater Dancer, but Springwater is rare here (note its distribution). Vivid has slightly shorter, stouter abdomen and cerci with down-turned points more tapering than Springwater. Female differs from most other female dancers in Rockies by combination of large size, moderate amount of striping on thorax, unforked shoulder stripe and pale abdomen tip. Reliably separating female Vivid and Springwater requires a microscope.

Distribution: All Rocky mountain regions in Foothills and Montane Zone.

Life Cycle & Behavior: Adults emerge in June and July and fly through September. Adults perch on bare ground, rocks or streamside vegetation. Males wait for females in sunlit areas near breeding sites and intercept them as they fly to water. Females lay eggs, usually in tandem, into plant stems slightly below the surface. Pairs stay in tandem for hours. Nymphs progress through 12 to 14 instars. Overwinters as a nymph. Populations have one- or two-year life cycles depending on stream temperature.

Sooty Dancer *Argia lugens*

APRIL	MAY	JUNE	JULY	AUG	SEPT	OCT

Breeding habitat: Small, rock-laden streams with open canopy.

Nature Notes:

Largest Rocky Mtn dancer.

A species of the desert and high plains.

Scientific name means "mourning," which may allude to its dark coloration.

The Sooty is so unlike other dancers that it was placed in a separate genus (*Hyponeura*) for many years. Dolly Gloyd, a prominent odonatist, synonymized the two genera in 1968.

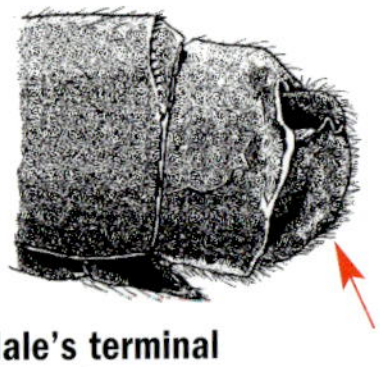

Male's terminal appendages.

Description: Adults average 1.8 inches long.

Large, dark-colored dancer. Maturing males with blue on the thorax quickly becoming covered with blackish pruinosity on top and sides (↑) shading to gray below. Dark shoulder stripes and dark top thorax stripe are both split for most of their length—visible on teneral males and females. His middle abdominal segments dark with tan apical rings; abdomen tip entirely dark. Female thorax either light brown or light blue like male. Her mid-abdominal segments like male, but abdomen tip is tan with dark side stripes. Wings of both sexes usually with amber wash, but may be clear.

Identification Clues: Both sexes have many distinctive features. Large size and overall pruinose black coloration of mature males is unmistakable. In the hand, the unbranched

This mature female is lunching on a leafhopper.

paraprocts (↑) easily separate it from all other dancers except Powdered, which has very different coloration. Split dark thorax top stripe and shoulder stripe easily distinguishes females and teneral males. Female large size and elongate mesostigmal lobes are also distinctive. Our Powdered and Sooty only dancers with two cells below the stigma of the wing, separated in the middle by a crossvein (both sexes).

Distribution: Foothills to lower Montane Zone of southern third of Southern Rockies.

Life Cycle & Behavior: Adults emerge primarily in June and July and may fly into September. Usually perches on rocks or sand. Mating usually takes place in early afternoon. Oviposition occurs in tandem, but with contact-guarding male not in sentinel position. Life history of this species has not been well studied.

Subarctic Bluet *Coenagrion interrogatum*

APRIL	MAY	JUNE	JULY	AUG	SEPT	OCT

Breeding habitat: Cold swamps and open bogs, especially quaking sphagnum moss margins around bog ponds.

Nature Notes:

One of the rarest damselflies in the Northern Rockies.

The southern limit of its distribution is further north than any other damselfly.

Found at Churchill on Hudson Bay and northwest to the Yukon Territories.

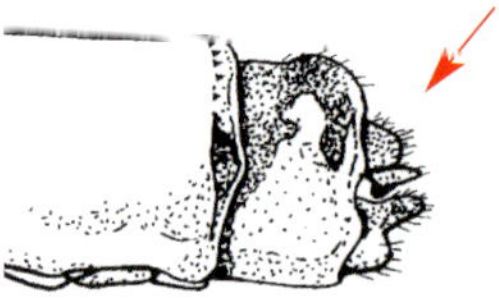

Male's terminal appendages: shape of cerci and paraprocts are unique.

Description: Adults average 1.2 inches long.

Both sexes are a study in blue and black. The pale shoulder stripes on the thorax are wide and bordered and divided with back, which forms a pale blue square and rectangle (↑) on the shoulder. The abdomen is blue and black, with the patterns on segment 2 unlike any other species. Pale eyespots are prominent on the otherwise dark head. Wings clear.

Identification Clues: Males have unique dark markings including the pale shoulder stripe feature described above and the underside of the thorax, which is extensively black including a Y-shaped figure (↑). The shape of the male's terminal appendages is diagnostic (↑). The female's thorax undersides are heavily marked with black and she shows long black streaks on the sides of her abdomen (↑).

Female Subarctic Bluets are heavily marked with black on the thorax and abdomen.

Distribution: Montane Zone at the northern tip of the Northern Rockies.

Life Cycle & Behavior: Flight season is early and relatively short, with emergence beginning in late May to early June. Females lay eggs in tandem into various leaves and stems near water level throughout June and perhaps into July. Sexual maturation takes about one week. Oviposition has not been described and the early nymphal stages are poorly known. Likely overwinters as a nymph in one of the last three instars.

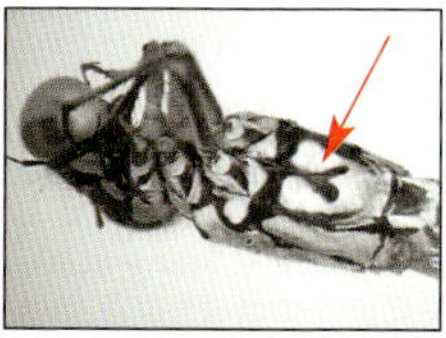

Male's distinctive Y-shaped mark on underside of thorax.

Taiga Bluet *Coenagrion resolutum*

APRIL	MAY	JUNE	JULY	AUG	SEPT	OCT

Breeding habitat: Wide variety of still waters from marshy ponds and grassy ditches to bogs, fens and slow streams.

Nature Notes:

One of the most abundant and widespread damselflies throughout the Rockies, including at elevations over 11,000 feet.

Nymphs overwinter embedded in ice but do not freeze solid!

Species name means "resolute," underscoring the fortitude they possess to eke out an existence in inhospitable northern climates—even above the Arctic Circle in Canada.

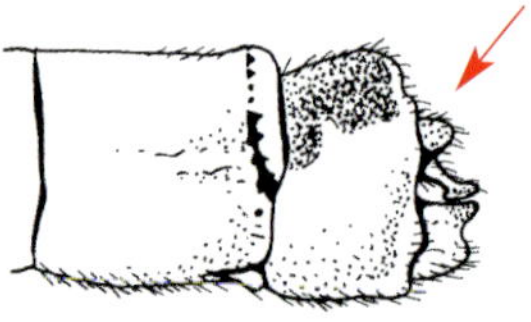

Male's terminal appendages.

Description: Adults average 1.2 inches long.

Attractive combination of blue, green and black. Male's thorax is blue with a wide black top stripe and thinner black shoulder stripes. Pale shoulder stripe may be divided into an exclamation point (uncommon). Sides of thorax blue, shading to green below. Abdomen is blue and black; segment 2 with a U-shaped dorsal black mark (↑). Segments 6 and 7 mostly black, 8 and 9 mostly blue. Female is similar, but with top of her abdomen almost entirely black, with very thin pale basal rings. Pale areas are either bluish green like male or light brown green. Wings clear.

Identification Clues: Identify males by blue-green color on side of thorax and U-shaped black mark on top of abdominal segment 2. Male has more black on abdomen than American bluets in the Rockies. Shape of male

Females resemble several American bluets but differ in lacking a vulvar spine on abdominal segment 8.

terminal appendages is diagnostic (↑). On females, look for blue-green sides on thorax, top of abdomen almost all black (↑) and the lack of a vulvar spine. Often perches with wings slightly spread (↑).

Distribution: All Rocky Mountain regions in Foothills, Montane and Subalpine zones. Found only at higher elevations in southern part of its range.

Life Cycle & Behavior: Adults emerge from late May to early July when water temperatures exceed 54 degrees F. Sexual maturation takes about one week. Pairing and egg laying start in June and adults fly into August. Mating occurs at the site of maturation, about 100 yards away from the pond, without courtship or territorial behavior. Oviposition is always in tandem and occurs below the surface into rushes, cattails, pondweed and other standing or floating aquatic plants. Eggs hatch within three weeks. Growth of nymphs is relatively slow compared to fast growers like spreadwing nymphs, extending for a period of ten to 22 months. Overwinters in nymphal diapause and embedded in ice in one of the last three instars.

Arroyo Bluet *Enallagma praevarum*

APRIL	MAY	JUNE	JULY	AUG	SEPT	OCT

Breeding habitat: Ponds, lakes and gentle streams with emergent vegetation; occasionally springs.

Nature Notes:

Rarely abundant like most pond damsels even in good habitat.

Females rarely seen; look for mating pairs.

Average life expectancy of adults is about five days according to an Arizona study.

Adults mate two or three times during their lifetime.

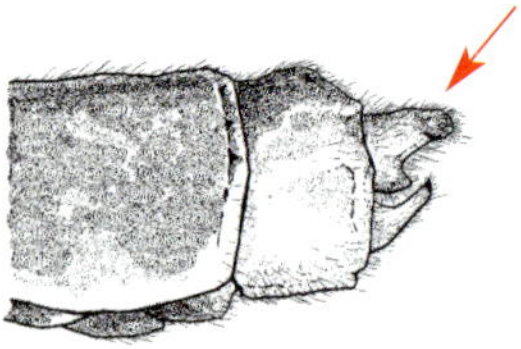

Male's terminal appendages: cerci distinctively shaped in profile.

Description: Adults average 1.2 inches long.

The male of this small bluet has a blue and black thorax with a thin dark shoulder stripe. Male's abdomen is much more black than blue on the middle abdominal segments but with conspicuous blue rings (↑); segments 6 and 7 nearly all black and segments 8 and 9 all blue. Female's pale areas are blue or light brown with thorax markings similar to male's. The top of her abdomen with segments mostly dark, with segments 3 to 6 pointed bullet-like toward her thorax creating pale markings at the bases of those segments.

Identification Clues: Male has more black on the middle abdominal segments than any other Rocky Mountain American bluet. Shape of his long cerci in side view is distinctive (↑) and easily separates from Tule Bluet male.

Females, like this blue-form example, are tough to identify without a microscope. This holds true for many American bluet females.

Male also similar to River Bluet but is a bit smaller, has more black than blue on the middle abdominal segments and cerci though similar in shape not as long. Female is similar to several other female bluets with narrow dark shoulder stripes, especially Familiar and Tule. Reliable identification requires examining the mesostigmal plates under a microscope. Identify females in the field by examining males with which they are mating.

Distribution: Parts of all Rocky Mountain regions, but primarily to the east of the Continental Divide in the Foothills and lower Montane Zone.

Life Cycle & Behavior: Adults emerge in June and July and fly well into August. Eggs are laid in tandem into floating vegetation or submerged plant stems, but male releases female when she submerges. Overwinters as nymphs, but nymphal stage not well known.

River Bluet *Enallagma anna*

APRIL	MAY	JUNE	JULY	AUG	SEPT	OCT

Breeding habitat: Small to mid-sized streams with gentle to moderate flow; sometimes warm springs and irrigation ditches.

Nature Notes:

A western species that appears to have recently expanded its range to the east.

May hybridize with Tule Bluet and possibly Familiar Bluet.

Carefully examine bluets along streams in the Rocky Mountains for this species.

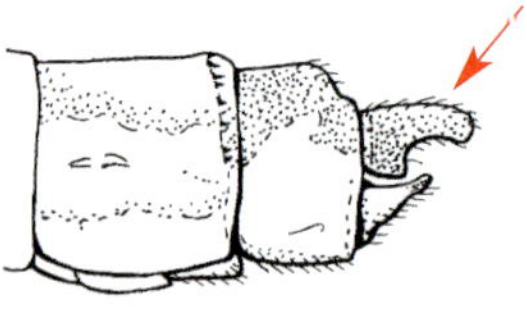

Male's terminal appendages: note long, distinctive cerci.

Description: Adults average 1.3 inches long.

The male of this running-water bluet has a blue and black thorax with a dark shoulder stripe narrowing rearward. Male's middle abdominal segments about half black, half blue, 6 and 7 mostly black and 8 and 9 all blue. His eyes black above, blue below. Female shows pale areas of blue, olive or brown, with narrow dark shoulder stripes. Dark markings on top of her abdomen torpedo-shaped, pointing toward thorax. Both sexes have eyespots of moderate size.

Identification Clues: Male is similar to other bluets with narrow dark stripes, but is readily separated from all in the hand—and sometimes in the field—by the long, distinctively shaped cerci (↑). Cerci of Arroyo Bluet similar to River in shape, but River's cerci

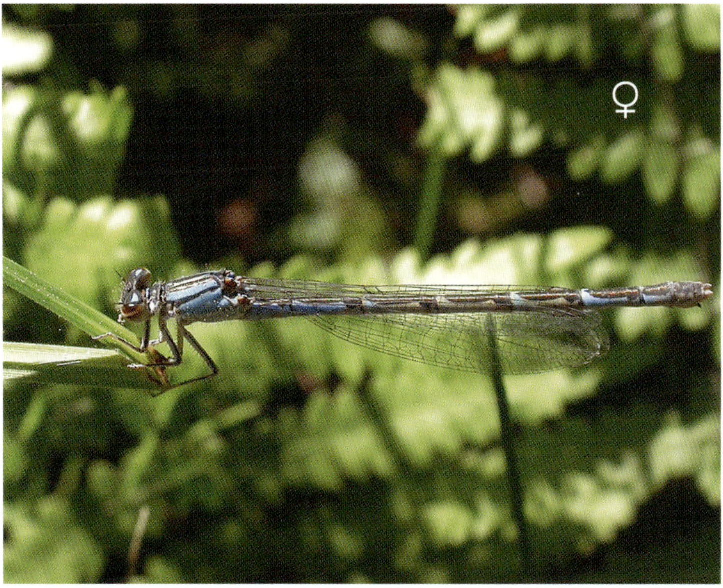

Females may be all blue, as this one, or partially or entirely olive or brown.

longer. River is also larger in size and has less black on middle abdominal segments than Arroyo. Female is very similar to Boreal and Northern Bluets and can only be identified with certainty by examining the mesostigmal plates under a microscope.

Distribution: Parts of all Rocky Mountain regions but most widely distributed in open country of Foothills and Montane Zone of Central Rockies.

Life Cycle & Behavior: Adults emerge in June and July and may fly into September. Adults frequently perch on the ground. Females oviposit singly or in tandem into floating vegetation and plant stems, often entirely underwater. The nymphs are a bit unusual among bluets in that they live in streams where they overwinter in later instars.

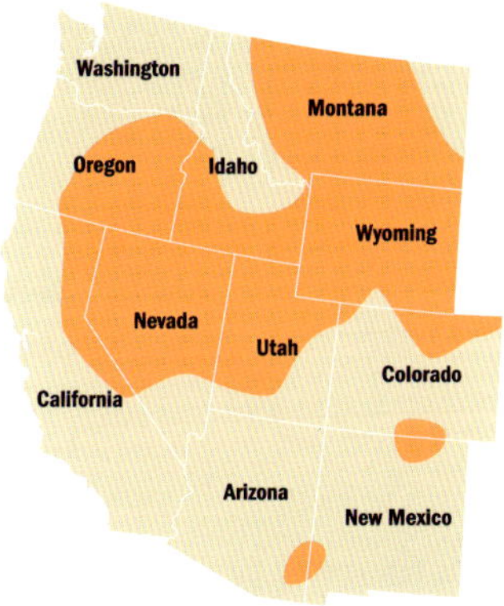

Familiar Bluet *Enallagma civile*

APRIL	MAY	JUNE	JULY	AUG	SEPT	OCT

Breeding habitat: Wide variety of still-water habitats including lakes, ponds, marshes and bogs. Even in brackish and alkaline waters.

Nature Notes:

This notable disperser is often the first damselfly to find newly created habitats like mitigation marshes and wildlife ponds.

Males are peaceful to a fault–they do not compete for territories, or court females, or engage in showy displays. Even the species name means "polite."

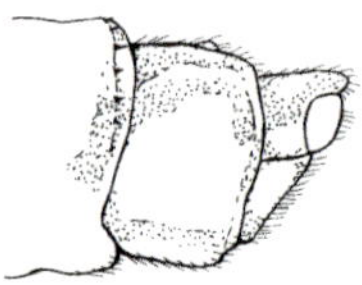

Male's terminal appendages: cerci have pale tubercles at tip.

Description: Adults average 1.3 inches long.

The male of this common bluet has a blue and black thorax with a fairly narrow, dark shoulder stripe. Male's abdomen is mostly blue on segments 2 through 5 (↑) with black areas increasing on 6 and 7; segments 8 and 9 are all blue. Eyes mostly blue. Female's pale area is blue to olive-tan with her thorax markings similar to male's. The top of her abdomen is mostly dark, but with small pale areas at the bases of some segments. Eyespots are moderate to small in size and tear-shaped.

Identification Clues: Male is a bit larger than the Marsh and Hagen's Bluet and usually has less black on his abdomen than the Tule, Alkali, River and Arroyo bluets. Northern and Boreal Bluets usually have larger eyespots. Cerci seen in side view are unmistakable with

This blue-form gal is another one of the many American bluet females difficult to identify without a microscope.

Mating pair in wheel formation

dorsal "arm" over a pale tubercle (↑). Female cannot be reliably distinguished from several other bluets except under a microscope.

Distribution: All Rocky Mountain regions but most common and widespread in Foothill and Montane Zone of Central and Southern Rockies.

Life Cycle & Behavior: The Familiar Bluet has a very long and apparently bi-modal, flight period, with emergence peaks in June or early July and again in late August and September. Some individuals may fly into October. Females oviposit into plant stems first in tandem above and at the surface and then alone beneath the surface. Nymphs live for nearly a year and overwinter in one of the later instars.

Tule Bluet *Enallagma carunculatum*

APRIL	MAY	JUNE	JULY	AUG	SEPT	OCT

Breeding habitat: Lakes, ponds and slow rivers, especially open shores of large lakes near bulrushes. Tolerates saline and brackish waters.

Nature Notes:

Often found at the open shores of larger, deeper lakes.

Tule Bluets have a long, mid-summer flight period in the Rockies.

Known to hybridize with Familiar Bluet and River Bluet.

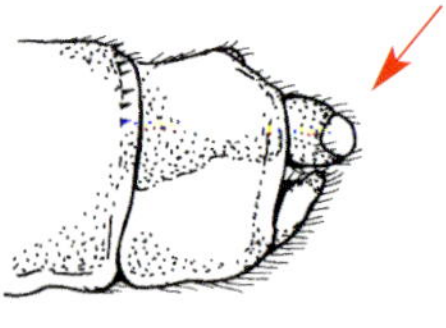

Male's terminal appendages: cerci have pale tubercles that extend out from tip.

Description: Adults average 1.3 inches long.

The male of this late-flying bluet has a blue and black thorax with a fairly narrow dark shoulder stripe of moderately even width. Male's abdomen is about half blue on segments 2 through 4 (↑) with increasing amounts of black on 5, 6 and 7; segments 8 and 9 are all blue. Female's pale areas are blue to olive-tan with her thorax markings similar to male's. The top of her abdomen is mostly dark but with small pale markings at the bases of some segments. Eyespots rather small, especially on the female (↑).

Identification Clues: Male shows more black on his abdomen than Familiar, Northern, Boreal, Marsh or Hagen's bluets and less black than Arroyo Bluet. Eyespots usually smaller than Northern and Boreal Bluets. Cerci seen in side view are unmistakable (↑)

Females come in several colors. Note the tiny postocular spots ("eye spots") on this blue-form female.

and only sure way to identify the male. Female cannot be reliably distinguished from several other bluets except under a microscope.

Distribution: Common in all Rocky Mountain regions in Foothills, Montane and Subalpine zones.

Life Cycle & Behavior: This species emerges in June and July and flies through September. Emergence may occur high on rush stems, a foot above water. Females oviposit in tandem into stems of bulrushes and other plants that may be submerged, near surface or well above water. Egg laying may occur into autumn. Nymphs live for nearly a year, have eight to 12 instars (usually ten) and overwinter in one of the later instars.

Alkali Bluet *Enallagma clausum*

APRIL	MAY	JUNE	JULY	AUG	SEPT	OCT

Breeding habitat: Alkaline and saline lakes, but not restricted to them; wave-swept shores of large lakes.

Nature Notes:

Flies alertly and low over open beaches and can be hard to catch.

Nymphs may crawl as far as 50 feet from water to emerge.

Alkali Bluet occurs at tremendous densities at preferred lakes due to reduced competition and predation.

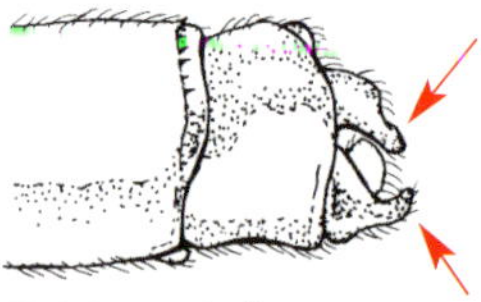

Male's terminal appendages: cerci are down-sloping. Paraprocts curve up at tip.

Description: Adults average 1.3 inches long.

The male of this enigmatic bluet has a blue and black thorax with a narrow dark shoulder stripe. Male's abdomen with increasing amounts of black on segments 2 through 7; segments 8 and 9 are all blue. His eyes mostly blue. Female's pale areas are blue or olive brown, with her thorax markings similar to the male's. The top of her abdomen is mostly dark, but with prominent pale basal rings and segment 8 entirely pale (↑). Eyespots of both sexes moderate in size.

Identification Clues: Similar to many other male bluets, especially River, Tule, Boreal and Northern. Has less black on abdomen than Arroyo Bluet and more black than Familiar Bluet. Terminal appendages seen in side view are diagnostic (↑) and only sure way to identify

Females are very similar to many other American bluet females but are the most likely of any to have abdominal segment 8 entirely pale.

the male. Cerci are short and downward-angled, somewhat similar to Northern Bluet with which I believe it has sometimes been confused. Females have larger pale areas on the abdomen than most bluets; especially look for the entirely pale segment 8 (but note that female Boreal Bluets can also have a pale segment 8). Firm identifications require seeing mesostigmal plates under a microscope.

Distribution: All Rocky Mountain regions in open areas of Foothills and Montane Zone.

Life Cycle & Behavior: Emerges in June and July and flies into September. Adults often alight dancer-like on the ground. Females oviposit in tandem into masses of filamentous algae and perhaps other vegetation that has collected along wave-lapped shorelines. Preferred habitat is too salty or alkaline for most odonates. Nymphs overwinter in one of the later instars.

Northern Bluet *Enallagma annexum*

APRIL	MAY	JUNE	JULY	AUG	SEPT	OCT

Breeding habitat: Marshy lakes, shallow ponds and slow streams. Needs fishless waters in the East but not in the West.

Nature Notes:

Occurs from coast to coast in North America, but is much more common in the West than the East.

A real high-altitude hero—the most common American bluet at high elevations in the Rockies.

Appears slightly later in season than very similar Boreal Bluet, but flight periods overlap.

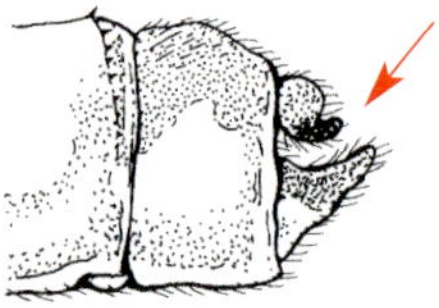

Male's terminal appendages: cerci have upturned tips

Description: Adults average 1.3 inches long.

Male of this mostly blue bluet has a blue and black thorax with narrow dark shoulder stripe. Male's abdomen has segments 2 through 5 mostly blue, 6 and 7 mostly black and segments 8 and 9 all blue. Eyes mostly blue, black on top. Female has pale areas blue or olive brown with thorax markings similar to male. The top of her abdomen is mostly dark with a pair of blue spots of variable size on top of segment 8. Both sexes have large eyespots; the male's nearly touch the eye (↑).

Identification Clues: Male is very similar to several other bluets, especially the Boreal Bluet, from which it can be separated only in the hand by viewing the upturned tip of cerci in side view (↑). Female can only be separated from other bluets with narrow, dark-shoulder

This blue-form female greatly resembles the male. More commonly, females will have an olive or brown abdomen and thorax.

stripes by examining mesostigmal plates under a microscope.

Distribution: All Rocky Mountain regions in Foothills, Montane and Subalpine zones. May creep up into the lower Alpine Zone.

Life Cycle & Behavior: Adults emerge primarily in June and July and fly into September. Females lay eggs in tandem into floating vegetation and submerged plant stems. Female may crawl a foot or more below the water's surface to lay eggs and stay submerged for 90 minutes. If the female gets caught in the surface tension upon reemerging, the male may "rescue" her by forming tandem and towing her to safety. Nymphs overwinter in one of the later instars.

Boreal Bluet *Enallagma boreale*

APRIL	MAY	JUNE	JULY	AUG	SEPT	OCT

Breeding habitat: Wide range of still-water habitats: shallow, well-vegetated ponds, marshy lakeshores, prairie sloughs and bogs.

Nature Notes:

Flight period early and short in East, extends through summer in Rockies.

A high-elevation regular, often occurring above 11,000 feet.

When a nymph loses a leg, it is able to regenerate a new nearly normal one.

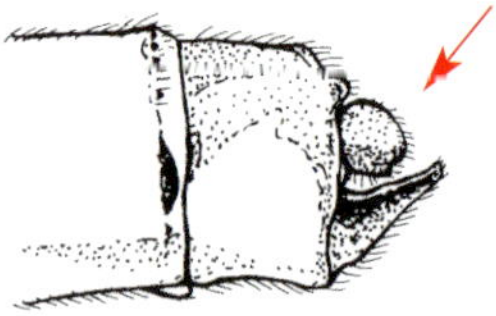

Male's terminal appendages: cerci rounded in side view.

Description: Adults average 1.3 inches long.

Male of this mostly blue bluet has a blue and black thorax with a dark shoulder stripe narrowing rearward. Male's abdomen has segments 2 through 5 mostly blue, 6 and 7 mostly black and segments 8 and 9 all blue. Eyes mostly blue, black on top. Female has pale areas blue or olive brown with thorax markings similar to male. The top of her abdomen is mostly dark with a pair of blue spots of variable size on top of segment 8 (↑) (often fused). Both sexes have fairly large eyespots; the male's nearly touch the eye.

Identification Clues: Male is very similar to several other bluets, especially the Northern Bluet, from which it can be separated only in the hand by viewing the rounded shape of cerci in side view (↑). Female can only be sep-

Female similar to other American bluets—especially Northern and Alkali—all of which have thin dark shoulder stripes, and can have extensive pale areas on top of abdominal segment 8.

arated from other bluets with narrow, dark-shoulder stripes by examining mesostigmal plates under a microscope.

Distribution: All Rocky Mountain regions in Foothills, Montane and Subalpine zones.

Life Cycle & Behavior: Emerges from late May through mid July and flies into September. Can be very abundant at preferred habitats. Females oviposit either in tandem or alone into emergent plant stems at or near the water's surface. Mating and egg-laying takes 1 to 1.5 hours. Nymphs overwinter in one of the later instars.

Marsh Bluet *Enallagma ebrium*

APRIL	MAY	JUNE	JULY	AUG	SEPT	OCT

Breeding habitat: Lakes, ponds, marshes and slow streams that have marshy borders. Most abundant in waters on calcareous soils.

Nature Notes:

Species name means "drunken," perhaps due to its occasionally erratic flight.

One female remained underwater laying eggs for five hours!

Virtually identical to Hagen's Bluet (except for male's cerci), but does not tolerate acidic conditions like that species does.

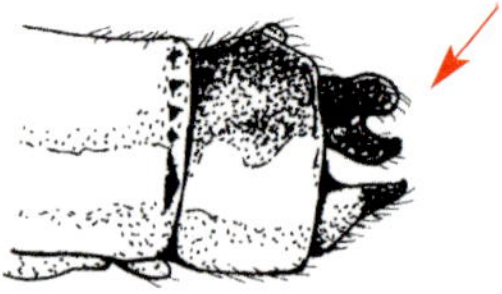

Male's terminal appendages: cerci with 2 arms equal in length.

Description: Adults average 1.2 inches long.

The male of this small bluet has a blue and black thorax with a dark shoulder stripe of moderate width. Male's abdomen is mostly blue on segments 2 through 5 with increasing areas of black on 6 and 7; segments 8 and 9 are all blue. Female's thorax is similar to male's, but may be light green, brown or even blue where he is blue. Her abdomen is all dark above (↑). Eyespots of both sexes moderate in size.

Identification Clues: Male is most like Hagen's bluet, but differs in having distinctive forked cerci with two arms about equal in length in side view (↑). Female Marsh and Hagen's Bluets differ from all other bluets by having the top of the abdomen entirely black, but can only be reliably separated from each

Top of female's abdomen is entirely dark on all segments, a trail shared only with Hagen's Bluet.

other by examining mesostigmal plates under a microscope.

Distribution: Foothills and Montane Zone of Northern and central Rockies.

Life Cycle & Behavior: Emerges in June over a three-week period and flies through the middle of August. Pairing and egg laying begin in late June and continue through July. Females oviposit alone or in tandem into floating plants—dead or alive—or plants stems below the surface, sometimes becoming fully submerged by a foot or more. Nymphs overwinter in one of the later instars.

Hagen's Bluet *Enallagma hageni*

APRIL	MAY	JUNE	JULY	AUG	SEPT	OCT

Breeding habitat: Well-vegetated ponds, lakes, marshes and bogs. Tolerant of acidic conditions.

Nature Notes:

A very common bluet when found in suitable habitat, but rare here.

Males are almost identical to the Marsh Bluet. Compare terminal appendages to distinguish them.

Males may pull reemerging females to safety after egg-laying, not altruistically but to mate with them!

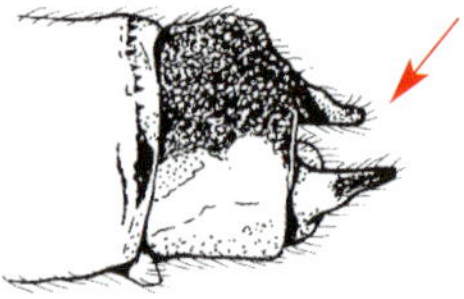

Male's terminal appendages: cerci have single horizontal arm.

Description: Adults average 1.2 inches long.

The male of this small bluet has a blue and black thorax with a dark shoulder stripe of moderate width. Male's abdomen is mostly blue on segments 2 through 5 with increasing areas of black on 6 and 7; segments 8 and 9 are all blue. Female's thorax is similar to male's, but may be light green, brown or even blue where he is blue. The top of her abdomen is entirely dark (↑). Eyespots of both sexes are moderate in size.

Identification Clues: Male is most like the Marsh Bluet, but differs in the shape of its cerci, which have a single horizontal arm seen in side view (↑). Female Marsh and Hagen's Bluets differ from all other bluets by having the top of the abdomen entirely black, but can only be reliably separated from each other

Top of female's abdomen is entirely dark on all segments, a trait shared only with Marsh Bluet.

by examining mesostigmal plates under a microscope.

Distribution: Rare in the Northern and Central Rockies where it is restricted to Foothills and lower Montane Zone on East side of Continental Divide.

Life Cycle & Behavior: Emerges in June and flies into September. Pairing and egg laying begin in late June and continue through July. Mating and egg-laying takes a pair about two hours. Females oviposit in tandem into floating plants, dead or alive or plant stems just below the surface, often without becoming fully submerged. Eggs hatch in about three weeks. Nymphs live for nearly a year and overwinter in one of the later instars.

Double-striped Bluet *Enallagma basidens*

APRIL	MAY	JUNE	JULY	AUG	SEPT	OCT

Breeding habitat: Lakes, ponds (may be temporary) and slow streams with emergent vegetation.

Nature Notes:

Like Great Spreadwing has expanded its range from U.S. Southwest to Northeast during the last century.

Often hovers over open water.

Abundant at suitable habitat, but rare here.

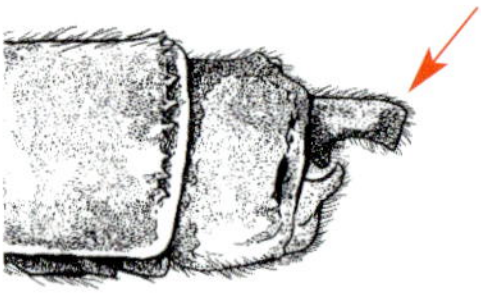

Male's terminal appendages: elongate upper arms of cerci extend far past short paraprocts.

Description: Adults average 1.0 inches long.

The male of this very small bluet is blue with black markings including dark top and shoulder stripes on thorax both divided into thin stripes along their entire length (↑). Males abdomen blue with black apical markings increasing in size from segments 3 through 7. Narrow black line on top of segment 3. Segments 8 and 9 all blue. Eyes entirely blue. Female is light brown on thorax with stripes as in male. Her abdomen all black on top, greenish blue on sides, except segments 9 and 10 which are light blue. Eyespots small and narrow with a pale bar between them.

Identification Clues: Tiny with unmistakable thorax stripes. Both sexes are unique among bluets in having the dark shoulder stripe narrowly divided along its entire length,

An exception to the rule, the female is easy to identify because of double thorax stripe, tiny size and pale blue abdomen tip.

with a thin pale stripe showing in between. Female's abdomen becomes increasingly blue toward the rear—unusual for a female American bluet.

Distribution: Rare in Foothills and lower Montane Zone of Southern Rockies on East side of Continental Divide.

Life Cycle & Behavior: Adults emerge primarily in late May and June and fly into September. Female lays eggs alone or in tandem into floating vegetation or slightly submerged plant stems. Nymphs overwinter in one the last instars. Life history of this species has not been well studied.

Desert Forktail *Ischnura barberi*

APRIL	MAY	JUNE	JULY	AUG	SEPT	OCT

Breeding habitat: Alkaline or saline lakes, ponds, marshes, ditches, canals and springs often in semi-arid areas.

Nature Notes:

Nymph stage is poorly known and the nymph has never been formally described.

Common at suitable habitats, where adults perch on grasses and sedges along shorelines.

Dennis Paulson reports that females oviposit alone, often laying eggs on the underside of floating leaves.

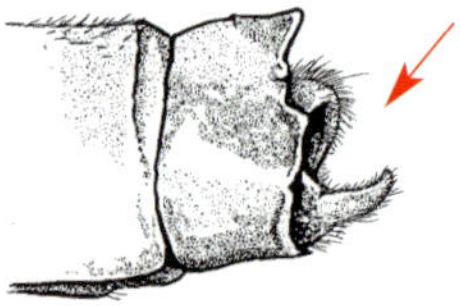

Male's terminal appendages: note extended paraprocts.

Description: Adults average 1.9 inches long.

A large, rather pale forktail. Male's pale shoulder stripe and lower thorax are light greenish blue (↑) and the tip of his abdomen is powder blue. His head is dark with blue-green eyes and small, blue eyespots connected by a blue bar. His abdomen black on top of segments 2 through 7 and light orange on the sides of segments 3 through 6 (↑). Fork-like protuberance on top of segment 10 is reduced. Female is highly variable in color. Male-like form occurs, but not common. Typical (heterochromatic) female is pale orange to brown with dark top and shoulder stripes thin or absent. Her abdomen is black on top, orange or brown on sides. Females turn dull brown all over with age, with some pruinosity on thorax.

Identification Clues: Male is unmistakable among our forktails due to combination of full

Note the lack of dark shoulder stripes on the thorax of this typical brown-form female.

pale shoulder stripes (not reduced to dots), golden-orange on sides of abdomen and lack of black marks on side of blue abdomen tip. Infer female identity by brown color, thin or absent thorax stripes (↑) and large size for forktail. Male-like females have same unique colors and patterns as males. Not all females are identifiable except under a microscope.

Distribution: Rare in Foothills and lower Montane Zone of Central Rockies in northern Utah and Southern Rockies in New Mexico.

Life Cycle & Behavior: Very little is known about the life history of this forktail. Evidently has a long spring and summer flight period and may persist well into autumn. Female lays eggs alone at water surface, often on underside of floating leaves. Overwinters as a nymph, probably in one of the later instars.

Pacific Forktail *Ischnura cervula*

APRIL	MAY	JUNE	JULY	AUG	SEPT	OCT

Breeding habitat: Wide variety of well-vegetated, still-water habitats: lakes, ponds, marshes, canals, ditches, warm springs and sluggish streams.

Nature Notes:

Mating occurs in the afternoon and lasts for hours; this prevents access to the female by other males until too late in the day.

Scientific name means "small deer," likely referring to the forked projection on top of segment 10 being reminiscent of a young buck's antlers.

Compare the male's larger forked projection with smaller one on Plains (pg. 124).

Description: Adults average 1.1 inches long.

A small, delightfully marked forktail. Male's thorax black with pale shoulder stripes interrupted to form four widely spaced blue dots (↑). Sides of his thorax blue. Abdomen is black (↑) except for segments 8 and 9 which are blue with black rectangles low on the sides. Forked projection on top of abdomen tip is pronounced (↑). Head dark with eyes black above, green below and small, blue eyespots. Females occur in a dizzying variety of colors and patterns. Typical-form has red eye spots and a pink thorax with thin, black top and shoulder stripes when young. With age thorax darkens to brown and is eventually obscured by dark gray pruinosity. Rare male-like female has thorax markings like male, becoming obscured with pruinosity. Her abdomen black above with mostly blue segment 8 and partly blue segment 9 regardless of age or form.

Striped thorax of female contrasts markedly with spotted thorax of male.

Identification Clues: Pacific and Plains males are the only Rocky Mountain forktails having four blue dots on top of black thorax. Separate them in the hand by differences in the terminal appendages and the much more elevated "fork" on top of segment 10 of Pacific (see illustration on previous page). Both female forms similar to Plains at same stages of development; they are reliably separated only with microscope.

Distribution: All Rocky Mountain regions in Foothills, Montane Zone and possibly Subalpine Zone.

Life Cycle & Behavior: Long flight period extends into October; adults begin to emerge in May and continue through summer. Newly emerged adults mature in just a few days. May perch with wings slightly spread, especially right after landing. Males perch in and near thickly vegetated areas near water. Female lays eggs alone into wide variety of vegetation at or slightly below the water surface. Nymphs have been found crawling on underside of lily pads. Overwinters as nymph.

Plains Forktail *Ischnura damula*

APRIL	MAY	JUNE	JULY	AUG	SEPT	OCT

Breeding habitat: Wide variety of well-vegetated, still-water habitats: lakes, ponds, ditches, warm/hot springs and sluggish streams.

Nature Notes:

Has a highly disjunct distribution in western Canada, with isolated populations occurring only at hot springs

Scientific name means "small deer," as does the scientific name of the Pacific Forktail, perhaps alluding to the similarity between these species.

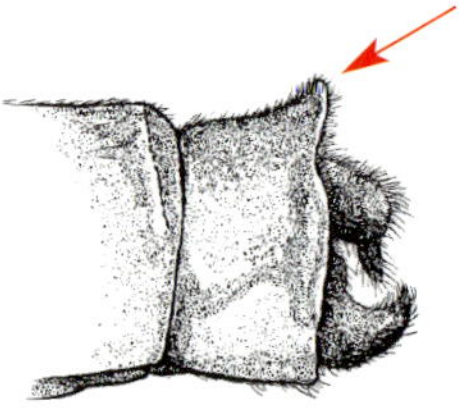

Compare the male's smaller forked projection with larger one on Pacific (pg. 122).

Description: Adults average 1.1 inches long.

Small attractive forktail with dotted thorax (↑). Male's pale shoulder stripes interrupted to form four widely spaced blue dots on his black thorax top. Sides of his thorax blue. Abdomen is black except for segments 8 and 9 which are blue with black rectangles on the sides. Forked projection on his abdomen tip not pronounced (↑). His head is dark with eyes black above, green below and small, blue eyespots. Females occur in a number of colors and patterns. Typical-form has pink eye spots (↑) and a pink thorax (↑) with thin, black top and shoulder stripes when young. With age thorax darkens to brown which is eventually obscured by dark gray pruinosity. Rare male-like female has thorax markings like male. Her abdomen black above with mostly blue segment 8 and often partly blue segment 9 regardless of age or form.

Note attractive combination of light pink and blue on this typical-form female. Inset shows rare blue-form female.

Identification Clues: Plains and Pacific males are the only Rocky Mountain forktails having four blue dots on top of black thorax. Separate them in the hand by differences in the terminal appendages and the little elevated "fork" on top of segment 10 of Plains (see illustration on previous page); Pacific has greatly elevated "fork." Typical and male-like females similar to Pacific at similar stages of development and are reliably separated by several differences on the prothorax seen under a microscope.

Distribution: Occurs throughout Southern Rockies in Foothills, Montane and Subalpine zones.

Life Cycle & Behavior: Fairly long flight period that extends into September; adults begin to emerge in May and continue into summer. Adults mature quickly after emergence. May perch with wings slightly spread, especially right after landing. Perches in thickly vegetated areas near water. Female lays eggs alone into aquatic vegetation at the water surface. Little is known about the life history of this species. Overwinters as nymph.

Mexican Forktail *Ischnura demorsa*

APRIL	MAY	JUNE	JULY	AUG	SEPT	OCT

Breeding habitat: Springs, slow streams and ditches; occasionally weedy shores of lakes.

Nature Notes:

Typical-form females live longer than male-like females, but also get harassed more by amorous males.

Like many forktails, female mates only once, stores that male's sperm in her sac-like bursa and uses it to fertilize all the eggs she lays throughout her adult life.

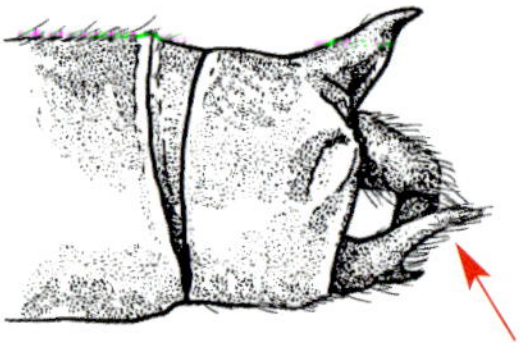

Upper fork of paraproct much longer than lower. Also compare to Western and Eastern Forktails.

Description: Adults average 0.9 inches long.

A tiny, handsome forktail. Male's thorax is green with black top and shoulder stripes. Abdomen is black above, light brown on sides of segments 3 to 6 and 8 and 9 are blue with black rectangles or streaks on the sides. His head is dark above with lower eyes green and small green eyespots. Younger female's thorax striped like male, but orange which extends onto base of abdomen. Rest of her abdomen mostly black. Females turn entirely bluish pruinose with age (just a few days), but the complete shoulder stripes may remain visible. Rare male-like female also becomes pruinose with age.

Identification Clues: Rectangular or streaky dark marks on sides of blue abdomen tip of males narrow possibilities to Mexican, Eastern and Western Forktails. Separate those three by shapes of forked paraprocts (primarily) and

Shoulder stripes are visible through the pruinosity of this mature female.

cerci (secondarily) in side view (see illustration on previous page). Mexican is unique in having upper fork of paraproct much longer than lower fork (↑). Also note restricted range, preference for streams and small size. Females of these three species can be identified with certainty only by examining the mesostigmal plates under a microscope.

Distribution: Parts of the Southern Rockies in Foothills and Montane Zone.

Life Cycle & Behavior: Adults have long flight period of April through October. Adults may be abundant in grasses and sedges around water at suitable habitat. Female matures in just a few days after emergence. Female mates only once and lays eggs alone into wide variety of floating vegetation or stems at water surface. Overwinters as nymph, but little else is known about nymph stage in Rockies.

Western Forktail *Ischnura perparva*

APRIL	MAY	JUNE	JULY	AUG	SEPT	OCT

Breeding habitat: Lakes, ponds and sluggish streams with grasses and sedges along shore; often at mud-bottomed, alkaline habitats.

Nature Notes:

Female brushes off unwanted suitors by fluttering her wings and curling her abdomen tip downward, in typical forktail fashion.

Not afraid to fly low and straight over the water, unusual for a forktail.

Scientific name means "very small."

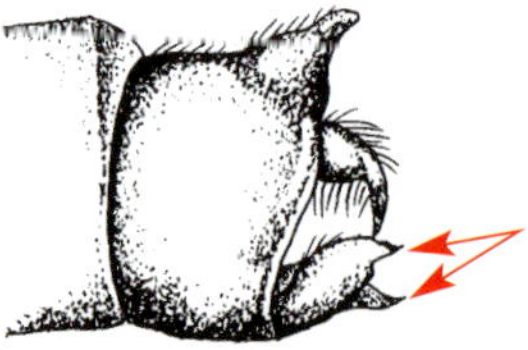

Upper and lower forks of paraproct about equal size. Compare to Eastern Forktail.

Description: Adults average 1.1 inches long.

A small, dark and strikingly marked species. Male's thorax is green with thick black top and shoulder stripes. Abdomen is black above from base through segment 7; segments 8 and 9 blue with black rectangles on the sides. His head is dark above with lower eyes green and small bluish green eyespots; face green. Younger female's thorax striped like male, but orange which extends onto base of abdomen and orange also appears in various amounts at tip. Rest of her abdomen black on top. Females turn entirely bluish or light gray pruinose with age (just a few days), but the complete shoulder stripes may remain visible. Very rare male-like female probably also becomes pruinose with age.

Identification Clues: Rectangular dark marks on sides of blue abdomen tip of males

This mature female's striped thorax is largely obscured by pruinosity.

narrow possibilities to Western, Mexican and Eastern Forktails. Separate those three by shapes of forked paraprocts (primarily) and cerci (secondarily) in side view (see illustration on previous page). Western is most like Eastern, but can be identified by its longer paraproct, with forks of about equal size (↑). Females of these three species can be identified with certainty only by examining the mesostigmal plates under a microscope.

Distribution: Widespread in all Rocky Mountain regions in Foothills, Montane Zone and possibly Subalpine Zone.

Life Cycle & Behavior: Long flight period that extends into October; adults begin to emerge in May and continue through summer. Adults mature in just a few days after emergence. Female mates only once and lays eggs alone into wide variety of vegetation at the water surface. Females frequently outnumber males at water. Overwinters as nymph.

Eastern Forktail *Ischnura verticalis*

APRIL	MAY	JUNE	JULY	AUG	SEPT	OCT

Breeding habitat: Wide array of permanent still waters including ponds, lake bays, marshes and slow streams.

Nature Notes:

Flies low over and through dense vegetation gleaning tiny insect prey off grass and sedge blades.

Female often flutters her wings at males, but only to show she is not receptive to his advances, not to encourage a relationship!

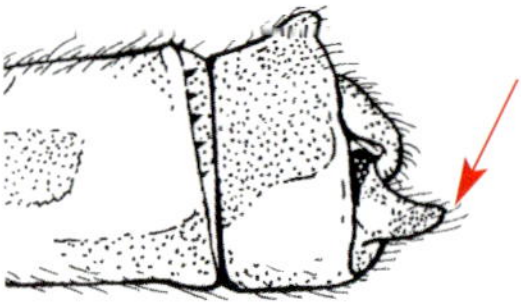

Lower arm of paraproct much longer than upper. Compare to Western Forktail.

Description: Adults average 1.1 inches long.

A small, dark twin to the Western Forktail. Male's thorax is green with thick black top and shoulder stripes. Abdomen is black above from base through segment 7; segments 8 and 9 blue with black rectangles on the sides. His head is dark above with lower eyes green and small bluish green eyespots; face green. Younger female's thorax striped like male, but orange which extends onto base of abdomen. Rest of her abdomen black on top. Females turn entirely bluish or light gray pruinose with age, but the complete shoulder stripes remain visible. Very rare male-like female probably also becomes pruinose with age.

Identification Clues: Rectangular dark marks on sides of blue abdomen tip of males narrow possibilities to Eastern, Western and Mexican Forktails. Separate those three by

Females come in two very different color varieties: immatures are black and orange, then turn bluish-gray pruinose with age.

shapes of forked paraprocts (primarily) and cerci (secondarily) in side view (see illustration on previous page). Eastern is most like Western, but can be identified by lower arm of the paraproct being much larger and longer than upper arm (↑); Arms of fork about equal in Western. Females of these three species can be identified for certain only by examining the mesostigmal plates under a microscope. Female Eastern has no orange at abdomen tip, female Western sometimes does.

Distribution: East slope of the Southern Rockies in Foothills and lower Montane Zone.

Life Cycle & Behavior: Adults begin to emerge in late May and continue to emerge throughout the summer. Some individuals may still be flying in October. Young adults mature quickly (males gain full color in one day) and do not fly far from the breeding site. Females may only mate once, storing that sperm for all future egg fertilization. Females oviposit alone (usually) into floating vegetation or submerged aquatic plants stems near water surface. Nymphs overwinter in various instars.

Black-fronted Forktail *Ischnura denticollis*

APRIL	MAY	JUNE	JULY	AUG	SEPT	OCT

Breeding habitat: Sedgy marshes, spring-fed slow streams and pools (hot or cold), often in semi-arid areas.

Nature Notes:

Only Rocky Mountain forktail that typically oviposits in tandem.

Adult is long-lived for a damselfly–about 6 weeks.

Males can be abundant at densely vegetated sites and look similar to the also tiny Sedge Sprite, so take a close look!

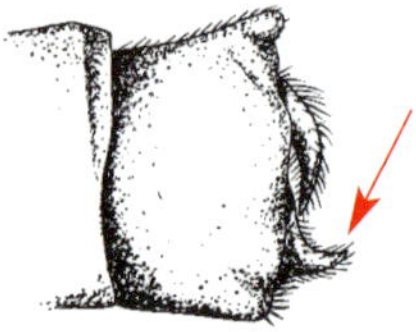

Note male's short cerci and paraprocts.

Description: Adults average 0.9 inches long.

A tiny black-topped forktail. Male's thorax shiny black above (↑) changing abruptly to bluish green on sides. No thoracic stripes. Abdomen is black above from base through segment 7; segments 8 and 9 blue with complete black streaks across sides creating blue top spots (↑). His head is black above with lower eyes green and tiny blue eyespots. Female colors and patterns highly variable. Typical female has black striped thorax over pink to tan base color when young; base color then becomes blue or green and eventually dark brown. Her abdomen dark above with some blue on segments 8 and 9, which becomes black with age. Also, a very rare male-like form exists.

Identification Clues: Male is unmistakable; only Rocky Mountain damselfly that is solid

Mature females are usually dark colored like this one.

black on thorax top with blue on segments 8 and 9 restricted to top spots rather than covering entire segments. Female resembles other forktails at various stages of her development but is smaller than Pacific or Plains and darker when mature than pruinose Eastern, Western, Mexican or Citrine.

Distribution: Central Rockies of Northern Utah including Uinta Mountains and southern half of Southern Rockies in Foothills and Montane Zone.

Life Cycle & Behavior: Adults have a long flight period, emerging primarily in June and July and flying into September or later. Adults require about a week to mature. Copulation is fairly short for a forktail, about 20 minutes. Female lays eggs in tandem in floating vegetation or moist plant stems near water surface. Nymph overwinters in a variety of instars.

Citrine Forktail *Ischnura hastata*

APRIL	MAY	JUNE	JULY	AUG	SEPT	OCT

Breeding habitat: Permanent but small ponds and spring seeps, often thickly overgrown with spike rush, club-rushes, grasses or sedges.

Nature Notes:

Called "exquisite little creatures" by pioneering odonatists Needham and Heywood.

One of the tiniest damselflies in the Southern Rockies and also one of the rarest.

Small size and cryptic coloration makes them hard to see in their densely-vegetated habitats.

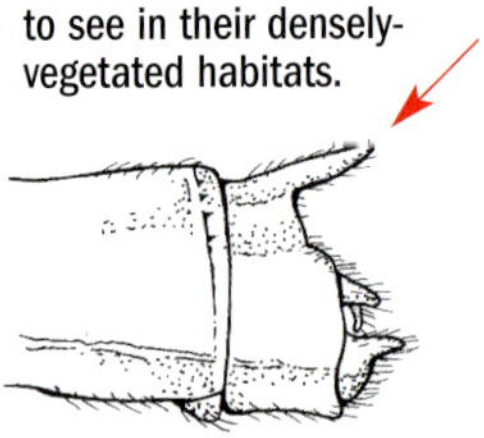

Side view of male's two-pronged fork on top of segment 10.

Description: Adults average 0.9 inches long.

Tiny yellow and black damselfly. Male's thorax is dark above and yellowish-green below with thin pale shoulder stripes. Abdomen is largely yellowish-orange with regularly spaced dark marks except at tip. When viewed from side, the tip of his abdomen shows a very prominent two-pronged fork (↑). Male has a dark top of the head with eyes green below and tiny blue-green eyespots. Young females are orange with dark top stripe on thorax and black on top of abdominal segments 6 through 9. Her dark shoulder stripe hairline thin or absent. Females darken with age, becoming increasingly bluish or gray with pruinosity.

Identification Clues: The male is easily recognized in the field by his small size, mostly yellow abdomen and prominent fork on top of

Mature females like this one are very similar to other pruinose female forktails.

segment 10. Orange stigma in forewing of male is unique in being well separated from the wing edge (↑). Look for the very thin or absent shoulder stripe on young (orange) females; older females are difficult to separate from other pruinose forktails although the lack of a dark shoulder stripe may remain evident.

Distribution: Rare in southern third of Southern Rockies in Foothills and lower Montane Zone.

Life Cycle & Behavior: Adult flight period is fairly long, from May into September. Females oviposit alone into stems of emergent vegetation. Females mate just once, storing sperm for future egg fertilization. Mating and oviposition are rarely seen and not well studied. Nymph likely overwinters in a variety of instars.

Painted Damsel *Hesperagrion heterodoxum*

APRIL	MAY	JUNE	JULY	AUG	SEPT	OCT

Breeding habitat: Small, gentle streams and ditches in arid to semi-arid areas.

Nature Notes:

Mature male is only red, blue and black damselfly in North America.

Unreceptive females flap wings with bright stigmas to discourage approaching males.

Like forktails, shy about being in the open, preferring to stay low in vegetation.

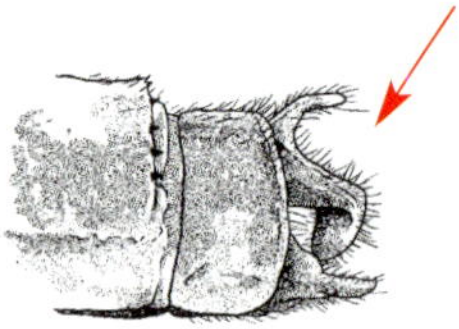

Male's terminal appendages showing uniquely S-shaped cerci.

Description: Adults average 1.3 inches long.

Mature male is a dazzling combination of red, blue and black. Thorax largely blue with black top stripe and thin black shoulder stripes interrupting pale shoulder stripes to form two large blue spots on each side that may be joined. Abdomen black above from base to segment 7; last three segments red (↑) with black rectangles on sides of 8 and 9. Stigma black with bright blue distal border. Eyes black above, green below, with eye spots large and red (↑). Immatures of both sexes start out orange-red and change colors dramatically as they age. Mature female colored much like male but without any red and top of segment 7 mostly blue.

Identification Clues: Mature male has unmistakable combination of red, blue and

Note extensive patch of blue on top abdominal segment 7 of this female.

black and uniquely shaped cerci (↑). Young male has same unique thorax pattern, but with distinctive red-orange body coloration. Female's extensive blue on top of abdominal segment 7 also unique. Very young females could be confused with Western Red Damsel females, but are more slender, have eye spots and lack "bump" under thorax.

Distribution: Scattered portions of Southern Rockies in Foothills and lower Montane Zone.

Life Cycle & Behavior: Adult flight period is long, from spring through autumn. Adults perch and forage fork-tail-like, low in dense beds of grasses and sedges. Males mate with females of all ages and colors. Females oviposit alone into a wide variety of aquatic plants, including watercress, near water surface. Nymph likely overwinters in a variety of instars.

Western Red Damsel *Amphiagrion abbreviatum*

APRIL	MAY	JUNE	JULY	AUG	SEPT	OCT

Breeding habitat: Gently flowing streams, ponds, marshes and seeps, often associated with springs (hot and cold), but not restricted to them.

Nature Notes:

Flies close to vegetation, avoiding open water.

Areas of preferred habitat may be small, but capable of producing many individuals.

High elevation not a problem; often found above 9,000 feet in Colorado.

Male's terminal appendages: thin cerci slope downward.

Description: Adults average 1.0 inches long.

Small, stocky, two-tone species. Male is mostly red (↑) with black areas that increase in size with age on top of the thorax and last three or four abdominal segments. Female is similar but duller, lacking black on top of her thorax. Her thorax and abdomen mostly brownish orange with reddish highlights on abdomen. Head and eyes reddish brown without eye spots. Wings clear. Both sexes with prominent, hairy "bump" (↑) on underside of thorax. Female has strong vulvar spine.

Identification Clues: Our only red and black damselfly that lacks eye spots. Downward sloping cerci of male seen in side view are diagnostic (↑). Look for lack of eye spots, "bump" on venter of thorax and no black areas on the female; other similar pond

Females undergo major color changes, darkening as they age.

damsels lack "bump" and that color combination.

Distribution: All Rocky Mountain regions in Foothills, Montane and Subalpine zones across many landscapes from arid to heavily forested.

Life Cycle & Behavior: Adults have long flight period emerging in May and June and flying into September, but more common in spring. Females oviposit in tandem, often with male in "sentinel" position, into wide variety of floating vegetation. Overwinters as nymph in one of later instars. Nymphs have been found in seepage areas with very little water.

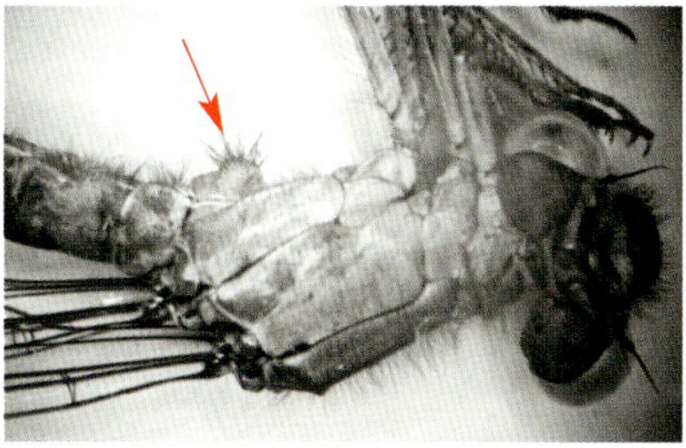

Note characteristic "bump" on underside of thorax.

Sedge Sprite *Nehalennia irene*

APRIL	MAY	JUNE	JULY	AUG	SEPT	OCT

Breeding habitat: Wide variety of habitats with abundant sedges including bogs, fens, marshes and pond and lake shores.

Nature Notes:

A locally common damselfly in the Rockies where habitat is suitable.

The female keeps her body horizontal when egg laying as the contact-guarding male supports himself at a 45 degree angle by holding her prothorax with his claspers.

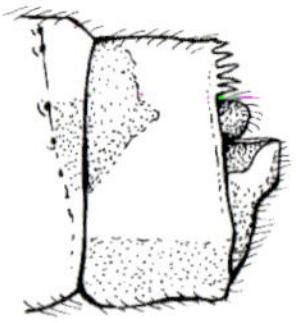

Male's terminal appendages.

Description: Adults average 1.0 inches long.

Very small (↑) metallic green and powder blue species. Thorax and abdomen are unstriped, metallic green on the top half (↑), light blue below. Abdominal segments 9 and 10 of male are pale blue with dark areas at bases of the segments. Eyes blue, face pale blue. No eyespots (↑), but a thin blue bar is usually present along the back of the head. Females like male but with less blue on segment 8. Vulvar spine lacking or tiny in female.

Identification Clues: Unmistakable due to their small size, unstriped metallic-green thorax and lack of eyespots, which sets them apart at a glance from all other Rocky Mountain damselflies.

Distribution: Northern part of Northern Rockies and Central and Southern Rockies in

Females are very similar to males but have less blue at the tip of the abdomen.

eastern Wyoming in Foothills and Montane Zone.

Life Cycle & Behavior: Adults emerge in late May or June and often fly throughout the summer. Females lay eggs in tandem in floating bits of dead rushes or other vegetation, floating or upright, or into moist sphagnum moss just below water surface. Nymph overwinters in one of the final two instars.

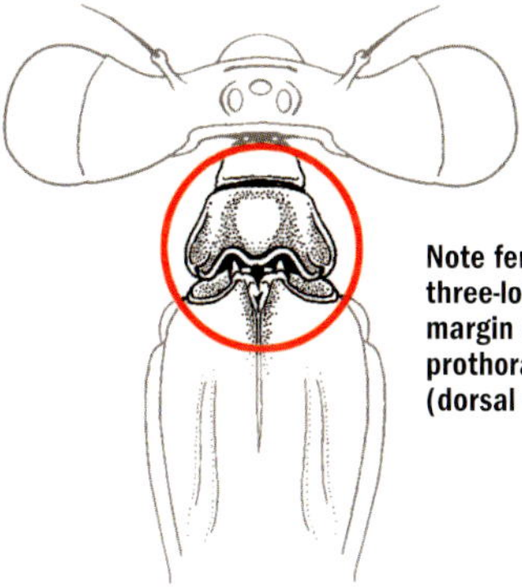

Note female's three-lobed hind margin of prothorax (dorsal view).

Darners
Family Aeshnidae

Darners are large, strong-flying dragonflies with huge eyes and slender abdomens. Any large, strong-flying dragonfly in the Rockies that shows blue or green on its body belongs to this family. The compound eyes of most species cover most of the head and broadly meet along an eye seam on top. The mouthparts are appropriately sized for a large insect, allowing the larger species to give you a bit of a bite when removing them from a net if you're not careful. However, they won't come after you to bite, won't bite if they land on you and the bite is harmless anyway. The thorax is large, powerful and either double-striped on the side or unmarked. The abdomen is long, narrowed at segment 3 of males and tapers to a pair of distinctive claspers used for holding the female by the back of the head while mating. Females are usually less brightly colored than the males, their abdomen is usually shorter and stouter and they have well-developed ovipositors (unusual among dragonflies). As you might guess, they use their egg-laying tools to insert eggs into holes they make in plant material, much like damselflies do. Females lay the eggs alone (except for Common Green Darner). Darners are primarily pond species and are ferocious predators in both adult and nymph stages. The great flight speed and visual acuity of adults allows them to home in on and devour smaller flying insects with astonishing rapidity. When at rest, most species hang vertically on vegetation.

Worldwide there are about 425 described species of darners, with about 41 species in North America north of Mexico. This is a well-represented family in the Rocky Mountains, with 13 species in three genera. Mosaic darners (genus *Aeshna*) are primarily a late-season (summer and autumn) group, while the Neotropical Darners (genus *Rhionaeschna*) emerge fairly early, and the Common Green Darner, which is a migrant, is usually the first odonate we see in the spring and one of the last to emerge in the fall.

Mosaic Darners — Genus *Aeshna*

Also known as the "blue darners," the lion's share of Rocky Mountain darners falls into this fascinating genus of very large to moderately large insects. Males are brown as base color with conspicuous pairs of blue, green or yellow stripes on the top and sides of the thorax. The abdomen is typically marked from base to tip with pairs of large blue spots dorsally and rows of smaller dots low on the sides. The face is pale, often with a dark cross line and there is a black T-shaped mark on

T-spot on frons, common to all mosaic darners.

top of the frons (forehead). Females have markings similar in shape to the males and each species usually has two or three color forms. Typically colored females (heteromorphs) usually have their stripes and spots yellow or yellowish green. All of our species also have a male-like (blue) form as well (andromorph) that may be fairly common or rare depending on species. The thorax is packed with powerful flight muscles needed to move their large wings. A number of our mosaic darners are found with some regularity in the Subalpine Zone where they are a prominent part of the community of aquatic predators. There has been much published speculation about what the name *Aeshna* means, but the bottom line is nobody knows for sure.

Mosaic darners are primarily lake, pond and marsh species where males can often be found patrolling along shorelines chasing other males and devouring copious quantities of smaller insect prey. In their gluttony, many species keep right on feeding through the afternoon and into the evening. They often congregate in large, mixed-species swarms when prey (mostly small flies) is abundant and they often fly high enough to be out of netting range. Once when frustrated from trying to net an especially high-flying swarm of mosaic darners, I balanced my net vertically on the tip of my index finger, and launched it about 20 feet straight up in air. When it landed there was a Shadow Darner flapping excitedly about in the bag. I try not to do this sort of thing too much with friends around because most of them lack my amazing netting skill and the last thing I want to do is make anyone jealous. Some species are also found at bogs, fens and streams. In fact, they fly so strongly that their travels could take them almost anywhere. They fly almost continuously, hovering occasionally or often depending on species, and when they do perch most species hang vertically from vegetation or perch vertically on tree trunks. This is primarily a late-season group (summer and autumn). Mating takes place in trees or bushes,

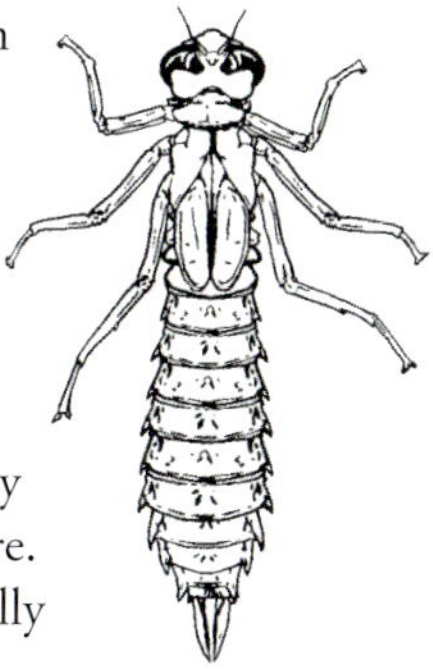

Typical nymph of the family Aeshnidae; They are long, cylindrical crawlers who stalk their prey amongst the plant litter and debris of the pond or stream bottom.

usually fairly close to water, and takes about an hour. Females of all species use their functional ovipositors to lay eggs into plant material. Nymphs grow to large size either quickly or relatively slowly depending on species, latitude, elevation and availability of food. Life cycle is usually two years or more; first winter typically in egg diapause and subsequent winters in various nymphal instars. The nymphs are large, slender and just as fiercely predatory as the adults. They are sometimes the top predators in the waters they inhabit, lurking on submerged vegetation to hunt smaller insects, tadpoles and even small fish.

Our species all have a pair of pale stripes on the side of the thorax, the shapes of which are useful in identification (see Thorax Comparison Chart, pg. 146-147). The abdomen is long, narrowed at segment 3 of males and tapers to a pair of claspers the shapes of which are also useful in identification. The cerci (upper claspers) fall into two basic types: paddle-type and wedge-type. Mosaic darners are not likely to be confused any other group except for the Neotropical Darners (genus *Rhionaeschna*). Species in these two genera can be challenging to identify to species because they are infrequently seen perched. But if flying specimens are netted and examined in the hand with a hand lens, the characters given in this guide will allow identification of all species. Since many species tend to fly a regular "beat," you can often set up an ambush to net one. I must repeat my familiar refrain: If you want to know what species of mosaic darner you are watching, you should net it and use a hand lens to look at the face, thorax stripes and terminal appendages, and other parts mentioned in this guide. This genus contains only about 42 named species worldwide, with 15 species currently recognized in North America and ten occurring in the Rocky Mountains.

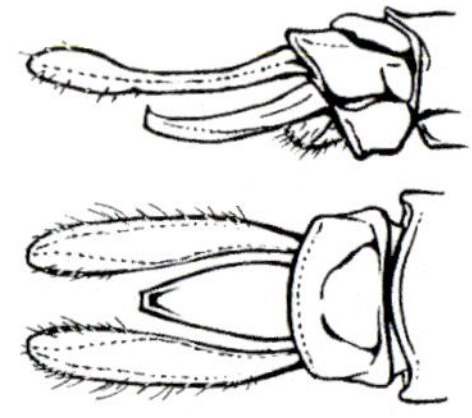

Paddle-type Claspers

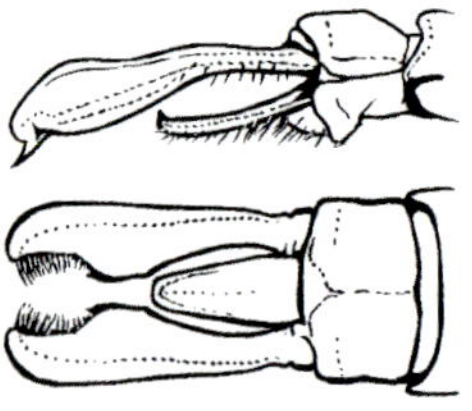

Wedge-type Claspers

Neotropical Darners — Genus *Rhionaeschna*

Rocky Mountain species in this attractive darner genus closely resemble the mosaic darners, from which this genus was only recently split. Differences between the genera include a noticeable "bump" (tubercle) on the underside or abdominal segment 1 (both sexes), and pale areas bordering the black T-spot on the frons (under the cap of the mushroom). Both characters are visible in the hand and are lacking on mosaic darners. Females come in two color forms: an uncommon male-like form (andromorph) with blue spots on the abdomen, and a typical, paler form (heteromorph) similar to mosaic darners. Worldwide there are 41 named species in this genus, with five found in North America. In the Rockies we have two species, one very common (Blue-eyed Darner) and one occasional (California Darner).

Green Darners—Genus *Anax*

This genus of large, strong-flying dragonflies is characterized by an unmarked thorax and many distinctive aspects of the wings including a long stigma. Adults are swift fliers that course over large areas of water or shoreline and do little hovering. *Anax* has been variously translated as meaning "lord," "master" or "ruler" so our single species, *Anax junius*, is quite literally "the lord (or master) of June!" Watch one and you'll understand why! Worldwide there are 28 named species in this genus, with five occurring in North America.

Identification of Mosaic Darners with Wedge-type Cerci

Structure	*Lance-tipped*	*Paddle-tailed*	*Shadow*
Black face cross line	*absent or very thin*	*moderate size*	*absent or very thin*
Underside of abdomen	*unmarked*	*unmarked*	*pale spots under segments 4-6*
Anterior thorax stripe	*widened at top*	*not widened at top*	*not widened at top*
Anterior thorax stripe	*clearly notched*	*may be slight notch*	*not notched, thinnest of group*
Anterior thorax stripe	*not black edged*	*may be black edged*	*black edged*
Female abdomen shape	*slender*	*stout, cylindrical*	*stout, cylindrical*
Female segment 9 (S9)	*as long as or longer than S8*	*shorter than S8*	*shorter than S8*
Female cerci	*oval, leaf-like, sharply pointed*	*elongate oval, bluntly pointed*	*elongate oval, bluntly pointed*
Rear of head behind eyes	*dark*	*dark*	*pale (tan)*

Mosaic & Neotropical Darner Thorax Comparisons

All *Aeshna* and *Rhionaeschna* species (the mosaic and neotropical darners) have two large pale side stripes on their thorax. Shape, color and width of these stripes is helpful in making a positive identification. But be aware that color is highly variable in some species. Refer to the species accounts for more field marks to look for when observing these darners.

We also show the male's terminal appendages in lateral view. These reproductive parts are often very distinct if you have them in the hand.

lateral thorax view

lateral male appendages

Black-tipped Darner *A. tuberculifera* (pg 148

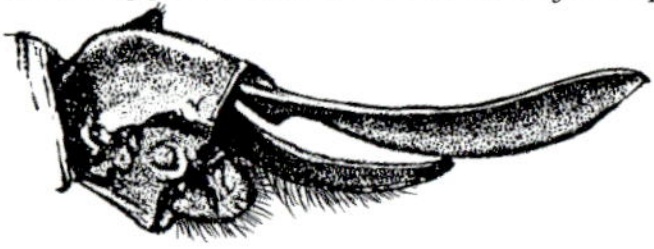

Sedge Darner *A. juncea* (pg 150)

Subarctic Darner *A. subarctica* (pg 152)

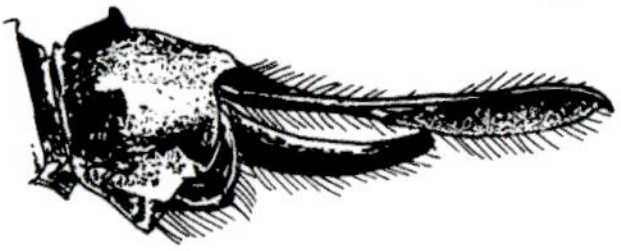

Canada Darner *A. canadensis* (pg 154)

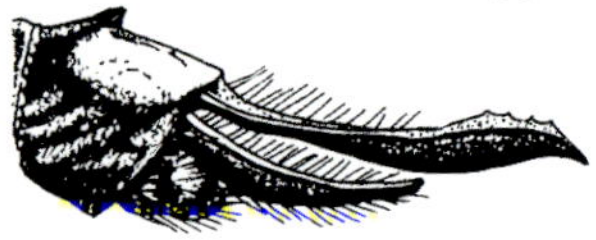

Lake Darner *A. eremita* (pg 156)

Variable Darner *A. interrupta* (pg 158)

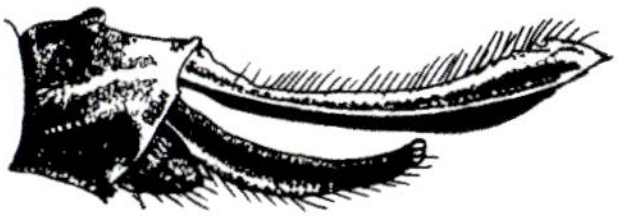

Zigzag Darner *A. sitchensis* (pg 160)

Lance-tipped Darner *A. constricta* (pg 162)

Paddle-tailed Darner *A. palmata* (pg 164)

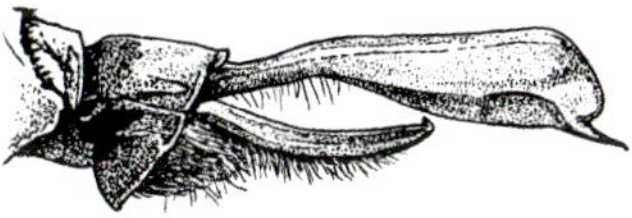

Shadow Darner *A. umbrosa* (pg 166)

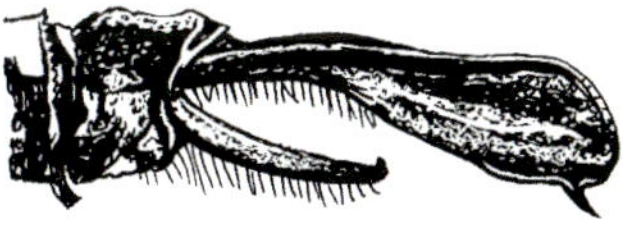

California Darner *R. californica* (pg 168)

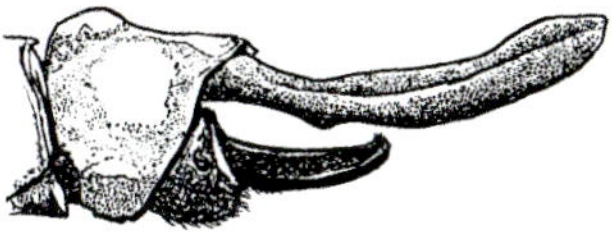

Blue-eyed Darner *R. multicolor* (pg 170)

Black-tipped Darner *Aeshna tuberculifera*

APRIL	MAY	JUNE	JULY	AUG	SEPT	OCT

Breeding habitat: Lakes and ponds in forested areas, often with bog vegetation; also well-vegetated slow streams.

Nature Notes:

Striking turquoise eyes and thorax stripes make this one of my favorite darners.

Male-like female form, which looks shockingly like male, is much more common than duller but usually typical female form.

Male-like look of female allows her to "fly under the radar" and avoid excessive harassment from males of other species.

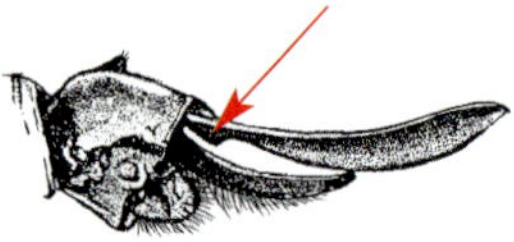

Note tubercle (small bump) near base of male cerci

Description: Adults average 3.0 inches long.

The male of this large, dark, slender species has a dark brown thorax and straight, parallel-sided, bluish side stripes (↑). Wasp-waisted abdomen is dark with dorsal spots blue except for segment 10, which is all black (↑). Claspers are paddle-type with a prominent knob near base of cerci (from which its name derives) (↑). Face pale (thin cross line rarely present), eyes turquoise, wings clear. Female usually looks very like male including coloration. Her abdomen more slender than most female darners. Rare heteromorph female has yellow-green spots. Female's ovipositor is large and cerci are long, pointed and very conspicuous.

Identification Clues: Straight-sided, blue thorax side stripes (see Thorax Chart pg. 146) and abdomen all black at tip (top of segment 10) easily separate both sexes from all other

This female is laying her eggs into a water-lily leaf.

darners. Male Shadow and Paddle-tailed Darners have wedge-type cerci. Both sexes have a distinctively large, dark, slender, narrow-waisted look.

Distribution: Uncommon in Foothills and lower Montane Zone of northern tips of Northern and Southern Rockies and east edge of Central Rockies.

Life Cycle & Behavior: Adults emerge primarily in July and fly into September. Flies high and fast, without hovering and often joins mixed-species feeding swarms. Female lays eggs late in the day into rush or sedge stems, or other forms of emergent vegetation, at or well above water surface; also into mud of drying pond bottoms. Usually has a two-or three-year life cycle with the nymphs progressing through about 13 to 15 instars. Spends first winter in egg diapause and subsequent winters as a nymph in various instars.

Note striking blue-green thorax stripes

Sedge Darner *Aeshna juncea*

APRIL	MAY	JUNE	JULY	AUG	SEPT	OCT

Breeding habitat: Small lakes, ponds, marshes, slow streams and fens with sedge margins in forested areas; prefers peaty waters.

Nature Notes:

Occurs above 10,000 feet in Colorado; most common darner in many high-elevation wetlands.

A mountain species in the West, not usually seen in desert, high plains or Foothills.

Scientific name *juncea* means "of the rushes," referring to its aquatic habitat.

Description: Adults average 2.7 inches long.

The male of this circumboreal species has a dark brown thorax and wide, straight, nearly parallel-sided side stripes that are blue above shading to yellowish-green below. Often a thin, pale streak between stripes (↑). Abdomen is dark with dorsal and lateral spots beyond segment 2 pale blue. Lateral spots on segments 1 and 2 yellowish green. Claspers are paddle-type. Face pale with thin black cross line, eyes blue or green and wings clear. Females may be male-like (blue-form) or more commonly, greenish yellow. Her abdomen stout and cylindrical with a small ovipositor and short cerci.

Identification Clues: Combination of very wide, nearly parallel-sided thoracic side stripes (see Thorax Chart pg. 146), thin black cross line on face and paddle-type claspers or small

Note the very wide side stripes with a thin strike between them on the thorax of this typical yellow form female.

ovipositor separates this species from our other darners. Also look for thin pale stripe between broad side stripes on thorax. Note females short cerci (shorter than segments 9 and 10 combined).

Distribution: All Rocky Mountain regions in Montane and Subalpine zones.

Life Cycle & Behavior: Adults emerge primarily in late June and July and may fly into October. Males patrol relatively small territories with slow, even flight interrupted with much hovering. Often hunts until dark. Female lays eggs late in the day into grasses, sedges, horsetails, matted roots or moss at or slightly below water surface. Nymphs progress through about 13 instars, which takes two years or more depending on latitude and elevation. Spends first winter in egg diapause and subsequent winters in various nymphal instars.

Subarctic Darner *Aeshna subarctica*

APRIL	MAY	JUNE	JULY	AUG	SEPT	OCT

Breeding habitat: Edges of mossy ponds and pools in bogs and fens; some northern swamps.

Nature Notes:

Hovers less and flies over larger territories than the more common Sedge Darner.

Needs ponds and pools in bogs with thick mats of Sphagnum mosses.

Note moderately thick face cross line.

Description: Adults average 2.7 inches long.

The male of this circumboreal species has a dark brown thorax and uniquely shaped side stripes bent forward in the upper halves. Side stripes are blue above shading to yellow below. Often with thin, pale lines between and in front of side stripes. Abdomen is dark with dorsal and lateral spots pale blue, sometimes shading to green and lateral spots yellow on basal segments. Claspers are paddle-type. Face pale with black cross line (↑), eyes green and wings usually clear but may be tinted brown on females. Females occur in three color forms: blue, green or yellow. Her abdomen stout and cylindrical with fairly long cerci.

Identification Clues: Thoracic side stripes bent forward in upper half are diagnostic (↑) (see Thorax Chart pg. 146). Also note moderately thick black face cross line. Most like

This green-form female is busy laying eggs into sphagnum moss.

Sedge Darner with which it may occur, but thorax side stripes much different, as are male hamular processes seen under a microscope. Frequent presence of thin pale lines on thorax both between and in front of side stripes gives thorax a more heavily striped appearance than most darners. Female has longer cerci than Sedge Darner (longer than segments 9 and 10 combined).

Distribution: Uncommon in Montane Zone of Northern Rockies.

Life Cycle & Behavior: Adults emerge primarily in July and fly into September. Males patrol territories only at low population densities. Flies over open bogs and fens at waist height or lower, with occasional hovering. Often hunts until dark and may join mixed-species feeding swarms. Female lays eggs into mosses and sedges at or just below water surface. Nymphs progress through about 12 instars, which takes at least two years, often longer in northern or high elevation habitats. Spends first winter in egg diapause and subsequent winters in various nymphal instars.

Canada Darner *Aeshna canadensis*

APRIL	MAY	JUNE	JULY	AUG	SEPT	OCT

Breeding habitat: Marshy and boggy lakes, ponds and slow streams in forested areas. Tolerates acidic conditions.

Nature Notes:

Male periodically drops down low amongst vegetation to search for females.

Foraging adults closely follow the contour of shoreline searching for prey around low vegetation in systematic manner.

Description: Adults average 2.7 inches long.

The male has a brown thorax and distinctive side stripes, typically blue above and green below, with yellow spot between them. The first (anterior) side stripe is deeply notched on its front margin (↑), narrowed above and with a rearward pointing offshoot (flag) at the top. Abdomen is dark with dorsal and lateral spots blue. Claspers are paddle-type. Face pale and unstriped with T-spot on frons (↑) — typical of all mosaic darners, eyes blue, wings clear. Typical females have pale areas green or greenish yellow. Rare male-like females have stripes and spots blue (see photo above, right). Her abdomen stout and cylindrical with cerci narrow and rounded at tip.

Identification Clues: Both sexes unmistakable because of deeply notched thoracic side

Captured in flight by a quick shutter finger, the deeply notched anterior side stripe of this male is readily seen.

stripe (see Thorax Chart pg. 146) and unstriped face. Slightly larger Lake Darner has side-stripe notch even more constricted and face has prominent black cross line. Male Lance-tipped Darner has much less pronounced notch on anterior side stripe and wedge-type cerci. Female Lance-tipped also has side stripe notch less pronounced than Canada and cerci are broad and pointed.

Distribution: Uncommon in Foothills and Montane Zone at northern tip of Northern Rockies.

Life Cycle & Behavior: Adults emerge in late June and July and fly through September. Male flies a territorial "beat" of about 60 feet along shoreline at waist height, hovering often. Often hunts until dark and may join mixed-species feeding swarms. Female lays eggs into mosses, algae or plant stems at or just below water surface. Nymphs progress through about 12 instars, which takes at least two years. Spends first winter in egg diapause and subsequent winters in various nymphal instars.

Lake Darner *Aeshna eremita*

APRIL	MAY	JUNE	JULY	AUG	SEPT	OCT

Breeding habitat: Wide habitat range including lakes, large ponds, occasionally slow streams in forested areas.

Nature Notes:

Threats abound everywhere, even for this large and powerful species. Eminent 20th century Canadian odonatist E. M. Walker records an egg-laying female's submerged abdomen being grabbed by a hungry nymph of her own species.

Scientific name *eremita* means "hermit," but not because it's reclusive. The species was discovered at Hermit Lake, NH.

Description: Adults average 3.1 inches long.

The largest mosaic darner in the Rockies. The male has a brown thorax and distinctive side stripes all blue or blue above, green below, often with pale spot or streak between them. The first (anterior) side stripe is deeply notched (pinched) on its front margin (↑), narrowed but rounded above and with offshoot at top often separated to a dot. Abdomen is dark with dorsal and lateral spots blue. Claspers are paddle-type. Face pale with a heavy black cross line (↑), eyes turquoise, wings clear. Typical females have pale areas greenish yellow. Rare male-like females have stripes and spots blue. Her abdomen stout and cylindrical with relatively large cerci.

Identification Clues: Deeply notched anterior side stripe (see Thorax Chart pg. 146) coupled with heavy black cross line on face and

Typical yellow-form female.

large size easily distinguish both sexes. Look for arched abdomen in flight.

Distribution: All Rocky Mountain regions in Montane and lower Subalpine zones.

Life Cycle & Behavior: Adult emergence begins in mid June and they fly into September. Male patrols tirelessly over water with slightly arched abdomen, but is not territorial. Flies about knee-height with little hovering. Hunts from dawn through dusk and may join mixed-species feeding swarms. Tough insect that can fly at low temperatures (50 F.) and in light rain. Often perches vertically on tree trunks; sometimes on ground. Female lays eggs into floating vegetation (including wood), root masses and emergent plant stems just below water surface. Nymphs progress through about 12 instars, which usually takes two years. Spends first winter in egg diapause and second winter in final instar.

Variable Darner *Aeshna interrupta interna*

APRIL	MAY	JUNE	JULY	AUG	SEPT	OCT

Breeding habitat: Marshy or boggy lakes and ponds with dense emergent vegetation; tolerates acidic waters.

Nature Notes:

The king of high altitude darners and one of the very highest altitude odonates! Breeding populations have been found very close to 12,000 feet in Colorado.

Several named subspecies of Variable Darner differ primarily in shape of thorax stripes and shape of male cerci; Rocky Mountain subspecies has narrow, uninterrupted first stripe, but intermediates are possible.

Description: Adults average 2.7 inches long.

The male of this mountain-loving subspecies has a brown thorax with reduced pale markings both on front (where sometimes absent) and on sides. Narrow, straight side stripes narrowed more near middle (↑), rarely completely interrupted forming upper and lower streaks. Side stripes usually blue above, yellow below. Abdomen is dark with dorsal and lateral spots large and blue. Claspers are paddle-type. Face pale with a thin cross line, eyes dark blue, wings clear. Females have pale colors either male-like blue (uncommon) or typical form yellow or greenish yellow. Her abdomen stout and cylindrical with long cerci.

Identification Clues: Both sexes can be identified by narrow, straight thoracic side stripes that may be interrupted in middle (see Thorax Chart pg. 147) and top thorax stripes

Uncommon blue form female.

tiny or absent. Thorax often looks dark in flight. Female most like Shadow, Paddle-tailed and Blue-eyed Darners. Lacks pale spots on underside of abdomen which female Shadow has. Female Blue-eyed has "bump" under abdominal segment 1. Variable female may not be distinguishable from Paddle-tailed except by examining the ovipositor under a microscope.

Distribution: All Rocky Mountain regions in Foothills, Montane and Subalpine zones.

Life Cycle & Behavior: Adults emerge primarily in June and July and fly into October. Males cruise shorelines with graceful flight, hovering often to look for females. Hunts until dark and often joins mixed-species feeding swarms sometimes on hilltops. Roams mountain clearings, sometimes far from water. Often perches vertically on tree trunks; sometimes on ground. Female lays eggs into floating vegetation (including wood), grasses, sedges and other emergent plant stems at or just below water level. Nymph stage typical for genus.

Zigzag Darner *Aeshna sitchensis*

APRIL	MAY	JUNE	JULY	AUG	SEPT	OCT

Breeding habitat: Small, shallow pools and puddles in bogs and fens with sedges and mosses; pools may dry during summer.

Nature Notes:

Usually perches horizontally, often on ground; unusual for a mosaic darner.

Captures some of its prey while prey is stationary on ground; also unusual for a darner.

Often rather tame and easy to approach, but not always so.

Note the zig-zag thorax stripes.

Description: Adults average 2.4 inches long.

The male of this small, boreal, peatland species has a brown thorax and distinctive thin side stripes bent into a zig-zag shape (↑) (hence its name). Front thorax stripes reduced or absent. All stripes light blue, sometimes interrupted and perhaps with a pale spot or streak between them. Abdomen is dark with dorsal and lateral spots large and heavily colored with blue. Claspers are paddle-type. Face pale with a thin black cross line, eyes brown with some blue and smaller than usual for mosaic darner, meeting along shorter seam. Typical females have pale areas blue or green on thorax, but yellow or white on abdomen. Rare male-like females have stripes and spots blue-gray including abdomen. Her abdomen stout and cylindrical.

This photo of a blue-form female shows the typical resting posture of this ground-perching species.

Identification Clues: Small size, zig-zag shaped thorax stripes (see Thorax Chart pg. 147) and abdomen heavily colored with blue are diagnostic. No other Rocky Mountain darner so marked.

Distribution: Northern and Central Rocky Mountain regions in Foothills, Montane and possibly lower Subalpine zones.

Life Cycle & Behavior: Adults emerge in late June and July and may fly into September. Flies low with more stops for perching than other darners. Female lays eggs near the edge of water into moss, algal mats or into mud. Nymphs progress through about 12 instars, which presumably takes two years. Spends first winter in egg diapause and second winter in one of the later instars. Nymphs survive under rocks only slightly damp when pools and puddles dry up.

Lance-tipped Darner *Aeshna constricta*

APRIL	MAY	JUNE	JULY	AUG	SEPT	OCT

Breeding habitat: Shallow, marshy ponds (may dry up in summer), lake edges and slow streams through open marshes.

Nature Notes:

Often perches (hangs) on low vegetation, unusual for mosaic darner.

Male-like form female patrols much like male, perhaps to avoid harassment by males of other species.

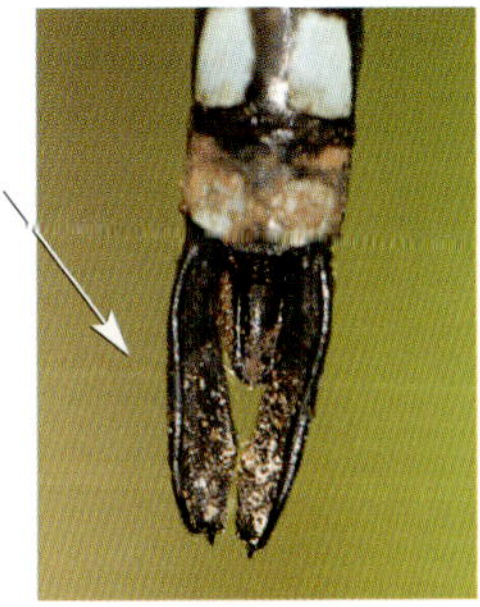

Male's wedge-type claspers. Dorsal view.

Description: Adults average 2.8 inches long.

The male has a brown thorax and distinctive side stripes blue above and yellowish green below. The first (anterior) side stripe is slightly notched on its front and rear margins (↑), widened and with narrow rearward offshoot at top. Abdomen is brown with large, blue dorsal and lateral spots. Claspers are wedge-type (↑). Face pale and unstriped. Typical females have pale areas yellowish or greenish yellow and wings often with orange tint on leading half. Rare male-like females have stripes and spots blue. Her abdomen more slender and more constricted at segment 3 than most darners (hence scientific name). Ovipositor very large for mosaic darner, thus segment 9 longer than in other darners. Her cerci large and lance-shaped (hence common name).

Identification Clues: Slightly but clearly notched anterior side stripe is unique (see Thorax Chart pg. 147). Other darners in Rockies have stripe deeply notched, straight or with barely perceptible notch or bent. Most often confused with Paddle-tailed and Shadow (see Table, pg. 144). Lacks pale spots on underside of abdomen. Note male's wedge-type cerci and lack of face stripe. Note female's slender abdomen; large ovipositor; long segment 9; and leaf-like, pointed cerci.

Distribution: All Rocky Mountain regions in Foothills and lower Montane Zone. Range may expand somewhat in wet years, contract in dry years.

Life Cycle & Behavior: Adult emergence begins in July and they fly into October. Male flies a slow territorial "beat" along shoreline at waist height, hovering infrequently. Hunts through dusk and may join mixed-species feeding swarms. Hangs on low vegetation (usually) or higher on branches. Female usually lays eggs high above water (up to waist height) in cattail or Sweetflag stems, but sometimes near water surface. Life history of nymph not well known; may grow more rapidly than most darners in ponds that periodically dry up.

Paddle-tailed Darner *Aeshna palmata*

APRIL	MAY	JUNE	JULY	AUG	SEPT	OCT

Breeding habitat: Wide variety of habitats; primarily partly shaded ponds, lakes and sluggish streams in forested areas.

Nature Notes:

Probably the most common Rocky Mountain mosaic darner.

Often flies in same habitat with very similar Shadow Darner, requiring continual checking.

I've found it at about 9,500 feet in Colorado.

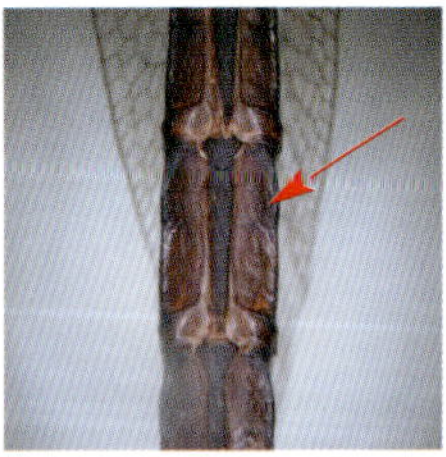

Underside of abdomen lacks pale spots. Contrast with Shadow Darner, pg. 166.

Description: Adults average 2.8 inches long.

The male has a brown thorax with side stripes nearly straight (↑) and often edged in black, colored blue-green above shading to yellow below. Abdomen is brown with large, blue dorsal and lateral spots. Claspers are wedge-type. Face pale usually with a thin but noticeable black cross line, eyes turquoise, wings slightly yellowed on females. Typical females have pale areas greenish yellow. Uncommon male-like females have blue spots on abdomen. Her abdomen stout and cylindrical with relatively large, bluntly pointed cerci.

Identification Clues: Shadow and Lance-tipped Darners similar to Paddle-tailed, all with wedge-type cerci (see Table pg. 144). Has noticeable black cross line on face; Shadow and Lance-tipped lack face cross lines or have only hint. Lacks pale markings on underside of

Typical green-form female in dorsal view.

middle abdominal segments, present on Shadow (↑). Lance-tipped has slight but noticeably notched anterior side stripe whereas Paddle-tailed has stripe straight or with tiny notch (see Thorax Chart pg. 147). Female Lance-tipped has abdomen more slender, with ovipositor larger, segment 9 longer and cerci more pointed.

Distribution: All Rocky Mountain regions in Foothills, Montane and Subalpine zones.

Life Cycle & Behavior: Adults emerge in late June and July and fly into October. Males patrol short shoreline areas, hovering often, until dark. Hunts low and slowly around edges of clearings. Often joins mixed-species feeding swarms. Female lays eggs into wet wood or into stems of various emergent plants up to waist height above water. Nymph stage not well studied.

Shadow Darner *Aeshna umbrosa occidentalis*

APRIL	MAY	JUNE	JULY	AUG	SEPT	OCT

Breeding habitat: Shaded forest streams, ditches and small rivers; also shaded lake shores and ponds.

Nature Notes:

Leaf-like cerci of female often break off when she oviposits into wood.

Scientific name *umbrosa* means "shadowy" or "shady"—either very apt.

Western subspecies has larger pale spots on abdomen than eastern subspecies.

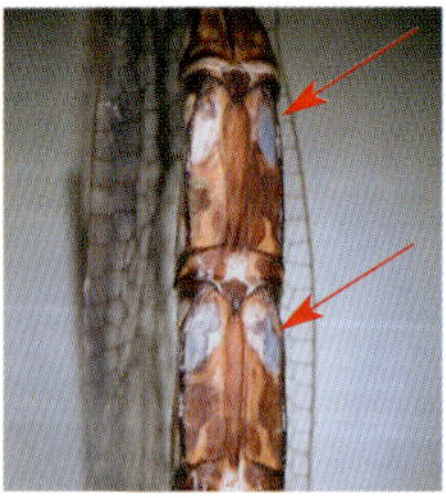

Underside of abdomen shows pale spots. Contrast with Paddle-tailed Darner, pg. 164.

Description: Adults average 2.8 inches long.

The male of this dark, shade-loving darner has a brown thorax with side stripes straight, thin and often edged in black (↑). Side stripes blue-green above shading to yellow below. Abdomen brown with relatively small, blue dorsal and lateral spots; segment 10 usually unmarked (both sexes). Pale spots under middle abdominal segments (↑). Claspers are wedge-type. Face pale with cross line absent or very thin, eyes turquoise, wings yellow-tinted becoming brown with age on females. Typical females have pale areas greenish yellow. Uncommon male-like females have blue spots on abdomen. Her abdomen stout and cylindrical with relatively large, bluntly pointed cerci.

Identification Clues: Shadow similar to Lance-tipped and Paddle-tailed, all with wedge-type cerci (see Table pg. 144). Best

Mating pair in wheel, perched in shady area.

Shadow character—very distinct pale spots on underside of abdominal segments 4 to 6. Also note absence of black cross line on face (or very thin); anterior side stripe thin and straight with no notch and often black edging (see Thorax Chart pg. 147) and spots on abdomen smaller than similar species. Female Lance-tipped has abdomen more slender, with pale spots larger, ovipositor larger, segment 9 longer and cerci more pointed.

Distribution: All Rocky Mountain regions in Foothills, Montane and Subalpine zones.

Life Cycle & Behavior: Adults of this primarily flowing-water species emerge in late June and July and fly well into October. Male patrols low and close to shore, hovering often. Hunts through dusk and may join mixed-species feeding swarms. Often perches vertically on tree trunks; sometimes on ground. Female lays eggs alone and late in day into wet decaying wood (rotting logs, stumps, branches), sometimes well above water. Nymphs progress through about 12 instars over at least two years.

California Darner *Rhionaeschna californica*

APRIL	MAY	JUNE	JULY	AUG	SEPT	OCT

Breeding habitat: Lakes, ponds, marshes and slow streams, sometimes alkaline, in forested or open landscapes.

Nature Notes:

One of the first Rocky Mountain dragonflies to appear in spring.

Pairs begin to mate while in flight, seek a place to alight, then continue for up to 20 minutes.

Usually perches vertically, low in vegetation.

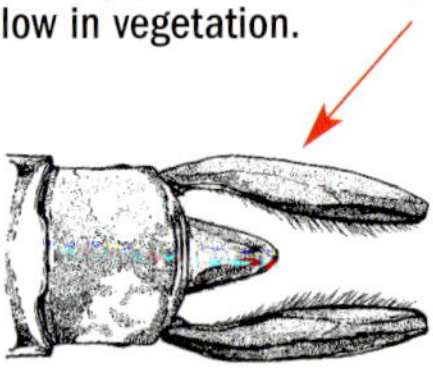

Dorsal view of male's abdomen tip. Note widely separated, bowed cerci.

Description: Adults average 2.4 inches long.

A small, early-appearing darner. The male has a brown thorax with light blue side stripes straight and narrow and often bordered with black behind. Top thorax stripes absent or reduced to tiny spots. Abdomen brown with dorsal and lateral spots large and blue (↑). Cerci similar to paddle-type of mosaic darners, but with small tubercle on ventral surface. Face blue with black cross line, eyes blue, wings clear. Typical females have pale areas yellow to white. Fairly common male-like females have stripes and spots blue; some intermediates. Her abdomen stout and cylindrical with long, narrow cerci.

Identification Clues: All other Rocky Mountain darners are considerably larger except for Zigzag, which has strongly bent, not straight, thoracic side stripes. Other darners

This typical-form female has a complex and interesting pattern of spots on her abdomen.

don't fly as early and most have top thorax stripes. "Bump" on underside of segment 1 and pale areas on sides of T-spot indicate genus. Black border only behind thorax side stripes also distinctive. Male smaller, earlier (though overlap), less brilliantly blue and with much different cerci than Blue-eyed. Females can usually be identified in the hand by combination of small size, thorax top and side stripe characters given above, black cross line on face and long slender cerci.

Distribution: Occasional in parts of all Rocky Mountain regions in Foothills and lower Montane Zone.

Life Cycle & Behavior: Adults emerge mostly at night in May and early June and fly into August. Males patrol low along shore in random fashion hovering occasionally to look for females. Female lays eggs alone into green plant stems at water level or into floating vegetation, including the underside of water lilies. Nymph stage not well-known, but early spring emergence implies overwintering in last nymphal instar.

Blue-eyed Darner *Rhionaeschna multicolor*

APRIL	MAY	JUNE	JULY	AUG	SEPT	OCT

Breeding habitat: Wide variety of marshy lakes, shallow ponds and slow streams in mostly open landscapes.

Nature Notes:

Brilliant blue spots on abdomen darken at cool temperatures.

Most common western darner in much of its range, but not above 7,000 feet.

Males have been reported to clash with, and drive off, low-flying Barn Swallows.

Note pale whitish areas on sides of T-spot.

Description: Adults average 2.7 inches long.

Male is dazzling, with brilliant blue eyes, face and body markings. Thorax is brown with blue side stripes straight and narrow (↑). Abdomen is brown with dorsal and lateral spots large and bright blue. Male cerci distinctively forked at tip. Typical females have brown eyes and pale areas yellow. Uncommon male-like females have eyes, face and abdomen spots blue, with side stripes blue-green. Her abdomen more slender than most female mosaic darners.

Identification Clues: Bright blue eyes and body markings of male are unmistakable; no other Rocky Mountain darner quite as bright. California Darner also bright blue, but not as bright as Blue-eyed and smaller. Forked tip shape of male cerci (pg. 147) diagnostic in west. Identify females by "bump" on underside

of segment 1 and pale areas on sides of T-spot (↑), which indicate genus. Female is noticeably larger than female California, is less likely to show a thin black cross line on face and lacks any hint of black edging behind pale thoracic side stripes.

Distribution: All Rocky Mountain regions in Foothills and Montane Zone.

Life Cycle & Behavior: Adults emerge mostly at night before midnight from late May through July and fly into October. Greatest abundances in July. Males patrol in random fashion over large areas of water at waist height, hovering intermittently, but may have short, defined "beats." Loves to course over open spaces—fields and roads—when feeding and may join mixed-species feeding swarms. Female lays eggs into wide range of vegetation types both above and below water surface. Thought to be migratory in part of its range (west of Rockies) and possibly elsewhere. Nymphs have been found tightly wedged in holes of submerged lava rock, but nymph stage not well-known.

Common Green Darner *Anax junius*

APRIL	MAY	JUNE	JULY	AUG	SEPT	OCT

Breeding habitat: Still, often shallow, marshy waters, including semi-permanent ones.

Nature Notes:

Our only darner that lays eggs in tandem.

Perhaps the most strongly migratory odonate in North America. Migrating adults from the south are usually the first odonates seen in spring, sometimes with ice not completely gone from ponds.

Perches near vertically, often on low vegetation, but does not hang like most darners.

Description: Adults average 3.0 inches long.

A large, uniquely colored darner, the male has an all-green thorax (↑) and abdomen with black stripe on top and wide, blue lateral stripes from segments 2 through 8. Segments 9 and 10 dark. Male cerci with spines at outer corners seen from above. Face unstriped, eyes greenish and a "bulls-eye"-like pattern (↑) on top of the blue frons. Rear of eyes edged in white or yellow. Typical female is similar to male but her abdomen reddish brown with gray-green lateral stripes. Male-like female also rarely occurs. Juveniles of both sexes have abdomen largely reddish brown (↑). Mature female's wings amber-tinted.

Identification Clues: Both sexes unmistakable. All-green thorax and "bulls-eye" mark on top of frons is diagnostic. Combination of features listed above not remotely approached by any other Rocky Mountain species, but some-

Prominent "bulls-eye" on frons is easily visible on this female.

what similar Giant Darner, which is larger and has a more slender abdomen, occurs not far to the south and west.

Distribution: All Rocky Mountain regions in Foothills and Montane Zone.

Life Cycle & Behavior: Migratory species in most areas, but also has a (smaller) non-migratory part of population. Adults fly into Rockies from areas to the south in April and May, with mating and oviposition occurring shortly after arrival. Nymphs have explosive growth rates in food-rich habitats, completing development and emerging in late summer and autumn. Shortly after, they fly south in huge numbers to reproduce. Their progeny are the adults that arrive from the south the following spring. Hunts from mid-morning to dusk, often swarming where prey is abundant. Males patrol territories over open water or along shorelines, flying tirelessly at about waist height. Females lay eggs in tandem with a guarding male in the afternoon—in the evening she may lay eggs alone. Eggs are laid into wide variety of standing plant stems and floating plant material including decaying wood.

Clubtails Family Gomphidae

There is something really special about the clubtails, but it's hard putting your finger on exactly what. Sure, their behaviors are interesting and they're attractive in their own modestly colored way. Many species are just rare enough to be challenging to find, and to heighten the challenge, their flight periods—our windows of opportunity to observe them—tend to be short. In fact, some species are rarely seen as adults even in prime habitat where they are common—as evidenced by gazillions of exuviae carpeting the banks. Then, to add to their mystique, many species live in aesthetically pleasing environments, like beautiful, undisturbed rivers and remote mountain streams. It is probably a combination of all of these things that makes them so compelling. Anyway, they're certainly near the top of my list of favorite odonate groups, and I'm far from alone among dragonfly enthusiasts in holding that opinion.

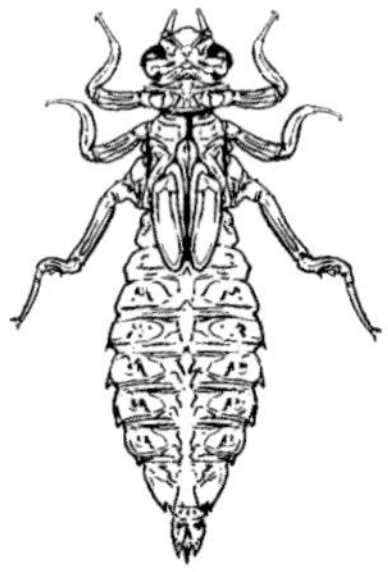

Typical nymph of the family Gomphidae; they demand water with a high oxygen content, which restricts most species to clean, faster-flowing rivers and streams.

The clubtails get their name from the widened "club" of expanded segments near the tip of the abdomen. Another signature feature of the group is that their eyes are well separated. The only other group of North American dragonflies with eyes that don't meet on top of the head is the petaltails (Family Petaluridae), but neither of the species in that family has been reliably reported from the Rockies. Clubtails are often colored with contrasting patterns of brown or black against yellow, or sometimes green, and most species (except for snaketails) have four or five dark stripes on each side of their thorax. Their wings are clear, with distinctive venation. Females are generally colored much like the males, but with a stouter abdomen and less of a club. Females have short subgenital plates instead of bladed ovipositors and lay their eggs without a guarding male nearby. All of our species breed in clean, well-oxygenated, flowing waters. They spend much of their time perched on the ground, rocks and other horizontal surfaces, from which they launch short forays to nab prey or chase off intruders. Most areas of North America are blessed with a nice diversity of species, but here in the Rockies we have a slim group of just five species in three genera, only one of which, the Pale Snaketail, is common throughout all the regions. Clubtail identification can be difficult (which probably adds to their mystique), usually requiring examining small reproductive parts with the specimen in hand. Our task here in the Rockies is simplified by the small number of species, but netting is still needed to identify all of them. Worldwide there are about 950 named species of clubtail, with 100 species in North America.

The Common Clubtails — Genus *Gomphus*

This is a diverse genus of moderately sized clubtails which currently serves as a "catch-all" group of sorts, certainly in need of taxonomic revision. The genus is divided into a handful of very useful subgenera from a practical viewpoint, but experts don't all agree about which ones are valid. Two of these subgenera are relevant to us in the Rockies. Members of the subgenus *Gomphurus* are characterized by an especially large club, relatively large overall size and long legs. Our single Rocky Mountain species of *Gomphurus* is the Plains Clubtail. Members of the subgenus *Gomphus* (recognized as *Phanogomphus* by some) are distinguished by a number of characters including the abdomen tip, which has a less prominent club than *Gomphurus.* Our single species in this subgenus, the Pronghorn Clubtail, has a club of moderate size.

The Hanging Clubtails — Genus *Stylurus*

These are late-season, mid-sized clubtails with slender abdomens and only a modest club on males. The genus usually has isolated, sausage-shaped pale stripes on top of thorax. However, our single species is an anomaly, having top stripes not isolated, but joining pale collar bar. Hanging clubtails sometimes perch horizontally like other clubtails, but get their name from their propensity to perch vertically, often caused by their landing on twigs or leaves that bend under their weight, thus they appear to hang.

The Snaketails — Genus *Ophiogomphus*

The snaketails are almost magical little creatures, flitting quick and low over riffles, appearing here, disappearing there. They inhabit clean, well-oxygenated rivers and streams, often in pristine landscapes, earning them the apt title of "the trout among dragonflies," applied to them by Sid Dunkle. This is a mid-sized group of clear-winged, short-legged species that have only a slight development of a club. They range in coloration from green to yellow with contrasting black or dark brown markings. A distinguishing feature of the genus is the nearly circular anal loop of the hind wing that usually contains three cells. The scien-

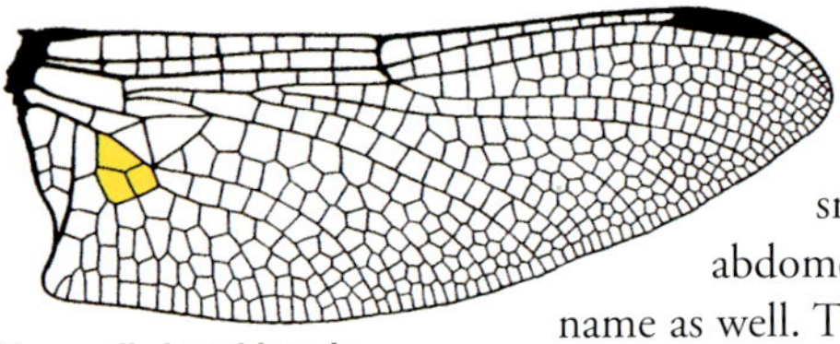

Three-celled anal loop in hind wing, characteristic of snaketails.

tific name of the genus essentially means "snake *Gomphus*" referring to snake-like markings on the abdomen, hence the common name as well. The nymphs are rather short and flattened with divergent wing pads and distinctive spatulate antennae. Nymphs are often abundant in good habitat and many species are very similar to each other, earning them the appellation "awfulgomphus," which I have used frequently when trying to identify them.

Dorsal view male terminal appendages | Lateral view male terminal appendages | Ventral female subgenital plate

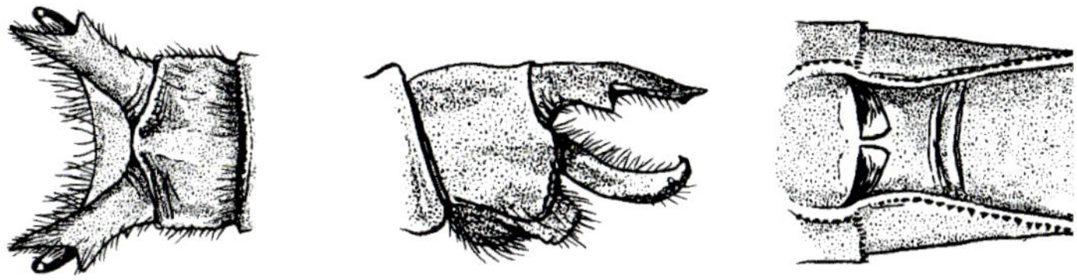

Pronghorn Clubtail (pg.178) *Gomphus graslinellus*

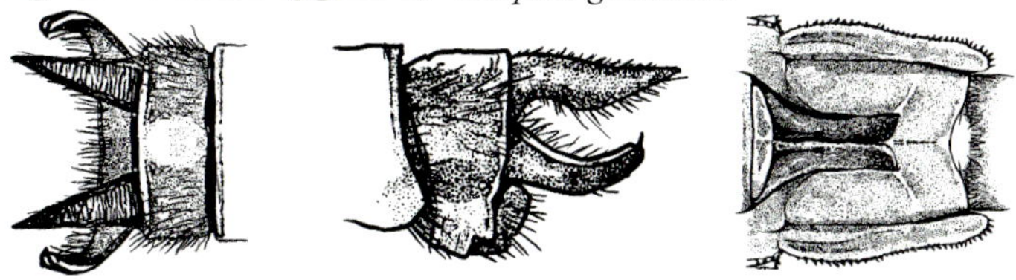

Plains Clubtail (pg. 180) *Gomphus externus*

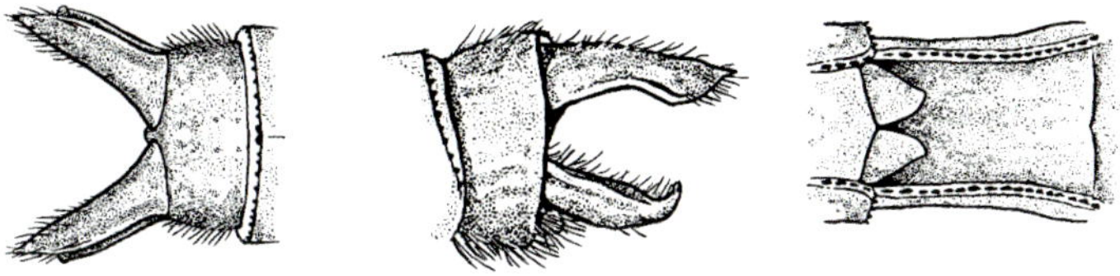

Brimstone Clubtail (pg. 182) *Stylurus intricatus*

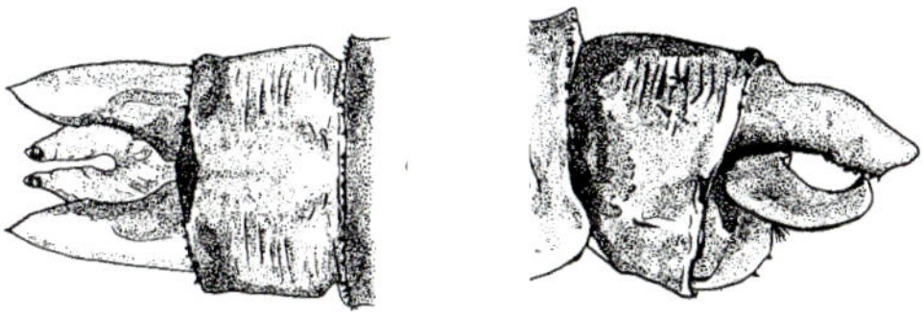

Pale Snaketail (pg. 184) *Ophiogomphus severus*

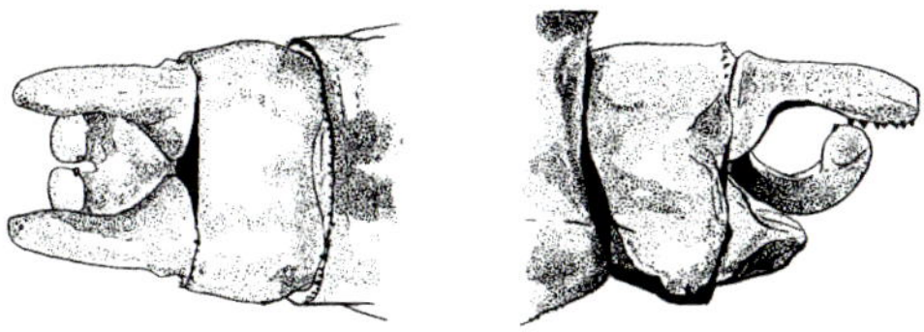

Sinuous Snaketail (pg. 186) *Ophiogomphus occidentis*

Pronghorn Clubtail *Gomphus graslinellus*

APRIL	MAY	JUNE	JULY	AUG	SEPT	OCT

Breeding habitat: Wide range of habitats including many types of streams, larger lakes and impoundments.

Nature Notes:

May obelisk to cool on hot days.

I've found this species to be rather wary and challenging to approach.

An eastern species rare here at the western edge of its range.

Common name alludes to the toothed shape of the male cerci in dorsal view resembling the horns of a Pronghorn Antelope (see below).

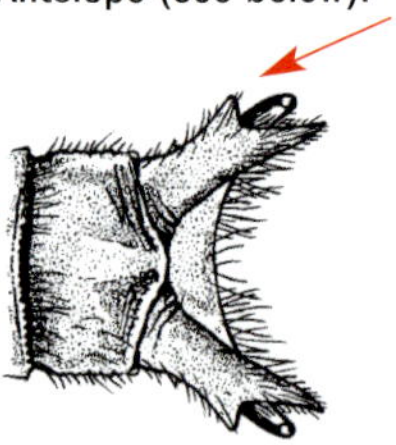

Description: Adults average 2.0 inches long.

A handsome and brightly colored species. Male thorax is yellow, heavily striped with dark brown or black. Abdomen is dark with yellow top stripes pointed toward rear on segments 1 to 8; yellow on segment 9 covering the entire top. Abdomen tip moderately clubbed with lateral margins of segments 8 and 9 yellow. Eyes blue, face pale and unstriped. Female similar to male in color, abdomen stouter and less clubbed. Her subgenital plate wide and short, about 1/6th as long as segment 9 and V-notched for half its length (pg. 177).

Identification Clues: Likely to be confused here only with very similar Plains Clubtail. Pronghorn is slightly smaller than Plains and lateral edges of segment 8 entirely yellow on Pronghorn (↑); some dark area there (may be tiny) on Plains. Those differences either not

Body colors of black and yellow form strong contrast on this species.

constant or slight, so cannot always be separated from Plains in the field. In the hand Pronghorn has prominent lateral tooth on male cercus lacking on Plains (↑) and the subgenital plate of female Plains is longer and has two pointed tips (pg. 177).

Distribution: Uncommon in the northern tip of the Northern Rockies and the eastern edge of the Southern Rockies in Colorado, in Foothills and lower Montane Zone.

Life Cycle & Behavior: Adults emerge in June and fly into August. Often perches on bare shore or on low plants. Adults usually fly in a series of dipping and rising concave loops when foraging, but can fly straight and fast. Less likely to hover over riffles than many other clubtails. Females lay eggs unattended near shore by flying slowly and descending at irregular intervals to tap the water surface with their abdomen, washing off dozens of eggs each time. Nymphs need at least two years to mature and spend final winter in last instar, thus ready to emerge in spring with favorable temperatures. Emerging nymphs crawl only a few inches out of water.

Plains Clubtail *Gomphus externus*

APRIL	MAY	JUNE	JULY	AUG	SEPT	OCT

Breeding habitat: Rivers and streams with moderate flow, often with muddy bottoms, sometimes sandy.

Nature Notes:

Perches alertly on open shores with abdomen slightly elevated, almost flag-like.

One scientist obtained 5,200 eggs from a single female by dipping her abdomen tip into a small dish of water.

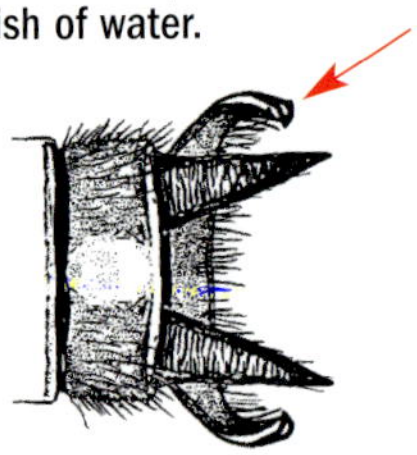

Male's cerci lacks lateral tooth of similar Pronghorn Clubtail. Also note wider epiproct.

Description: Adults average 2.2 inches long.

Colored like a larger version of the Pronghorn Clubtail, the male's thorax is yellow, heavily striped with dark brown or black. Abdomen is dark with yellow top stripes pointed toward rear on segments 1 to 8; yellow on segment 9 covering the entire top. Abdomen tip moderately clubbed with lateral margins of segments 8 and 9 mostly yellow. Eyes blue, face pale and unstriped. Female similar to male in coloration with a stouter abdomen and less club development. Her subgenital plate double pointed and long, over half the length of segment 9 (pg. 177).

Identification Clues: Likely to be confused here only with the very similar but slightly smaller Pronghorn Clubtail. See that species account for field character differences. Cannot always be separated from Pronghorn in the field. In the hand, Plains lacks lateral tooth on

Female is much like the male, but with a thicker abdomen.

male cerci, present on Pronghorn and epiproct is wider than Pronghorns (↑). Subgenital plate of female Plains is longer and more pointed than Pronghorns (pg. 177).

Distribution: Uncommon on both sides of Continental Divide in Foothills and Montane Zone of western edge of Central Rockies and eastern edge of Southern Rockies.

Life Cycle & Behavior: Adults emerge in June and fly through July. Males often rest alertly on open shores between intervals of chasing each other over open water. Female lays eggs unattended by descending while flying to tap the water surface with her abdomen to wash off eggs. Nymphs need at least two years to mature and spend final winter in last instar, thus ready to emerge in spring with favorable temperatures. Emerging nymphs crawl a foot or two out of water to transform.

Brimstone Clubtail *Stylurus intricatus*

APRIL	MAY	JUNE	JULY	AUG	SEPT	OCT

Breeding habitat: Warm, muddy, slow rivers and sometimes irrigation ditches, in open country, often arid.

Nature Notes:

The smallest and palest clubtail in the Rockies, but an attractive one.

Perches on low vegetation, but rarely on the ground.

Common name alludes to yellow coloration and its hot habitat.

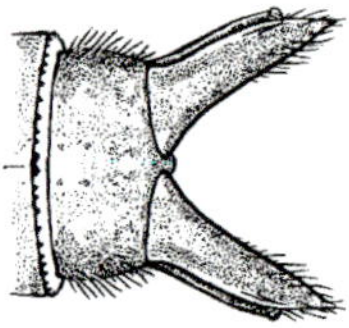

Dorsal view of male claspers.

Description: Adults average 1.7 inches long.

A small, pale, western clubtail with a club of moderate size. Male is mostly yellow (↑) with dark stripes on thorax and dark streaks and rings on abdomen forming a strong contrast. Eyes light blue and face pale. Female colored like male, sometimes with a bit more black on middle segments of abdomen. Her abdomen stout with little development of a club and her subgenital plate is short (less than half the length of segment 9) and split in the middle forming two points (pg. 177).

Identification Clues: Our only yellow clubtail. Likely to be confused only with Pale Snaketail, with which it often occurs. Brimstone is smaller than Pale and lacks green on thorax usually present on that species. In the hand, Brimstone hind wing lacks three-celled anal loop characteristic of snaketails (pg. 176).

Female is a small, attractive, mostly yellow clubtail.

Distribution: Rare in Foothills and possibly lower Montane Zone of southern tip of Central Rockies (including lower Uintas) and western edge of Southern Rockies.

Life Cycle & Behavior: Adults emerge in late June or July and fly into September. Males rest for lengthy periods, but are energetic in flight, flying low and swiftly over short beats. Pairs engage in lengthy copulation of an hour or more. Female lays eggs unattended by descending while flying low and swiftly to tap the water surface with her abdomen, washing off eggs. Little is known about the life history of the nymph.

Pale Snaketail *Ophiogomphus severus*

Breeding habitat: Streams and rivers with moderate current in open or forested areas; occasionally large, sandy lakes.

Nature Notes:

The only common clubtail throughout the Rockies.

Two subspecies have been named, both occurring in the Rockies. The form characteristic of the Northern Rockies has more extensive black markings.

Scientific name means "severe," perhaps referring to the arid landscapes where often found.

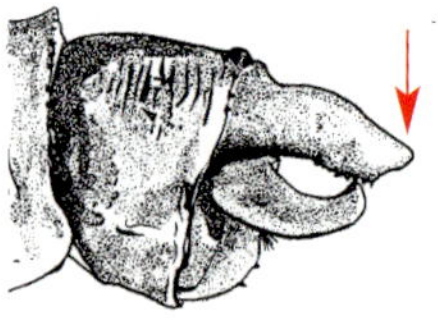

Note pointed tip of male cerci in side view.

Description: Adults average 2.0 inches long.

Male's thorax is pale green with dark shoulder stripes reduced or absent. If present, 1st shoulder stripe a dark oval and 2nd stripe a fine line in suture. Abdomen is pale green near base quickly grading to yellow with dark triangles on sides. Abdomen tip slightly clubbed. Eyes blue-gray; face pale, unstriped green. Female similar to male in color, but with a stouter, less clubbed abdomen.

Identification Clues: Pale Snaketail is larger than Brimstone Clubtail and has greener thorax. Western Pondhawk can look similar, but is not a clubtail, its eyes meeting on top of head. Snaketails have three-celled anal loop (pg. 176); all other groups have anal loop different. Paler than other snaketails. Sinuous Snaketail differs in having curved, slightly separated 1st and 2nd shoulder stripes on greener thorax, more black

Female usually shows green, not brown on her thorax and abdomen.

on abdomen and differently shaped male claspers with tips of cerci blunt (pointed on Pale (↑) —see also pg. 177). Female lacks small spine-like horns behind eyes or on rear of head, present on many snaketails (called occipital- and postoccipital horns).

Distribution: All Rocky Mountain regions in Foothills and Montane Zone.

Life Cycle & Behavior: Adults emerge in late June or July and fly through August. Males rest for lengthy periods, launching predatory strikes on small flying prey from ground, or rocks, occasionally low shrubs near riffles. Males are well known to obelisk at mid-day; flying long patrols when cooler. Most often seen in afternoon. Pairs engage in lengthy copulation. Female lays eggs unattended by making short foray from perch to tap the water surface with her abdomen, washing off eggs. She may only tap once, return to perch, then repeat. Nymphs likely take three years to develop and climb only about 6 inches from water to emerge.

Sinuous Snaketail *Ophiogomphus occidentis*

APRIL	MAY	JUNE	JULY	AUG	SEPT	OCT

Breeding habitat: Rivers and streams of moderate to slow flow, with sand or gravel bottoms, in open or forested areas.

Nature Notes:

A wholly western species (*occidentis* means western) confined to the northwestern U.S. and adjacent areas of British Columbia.

Female has distinctive behavior pattern of occasionally plunging full-length into water when ovipositing.

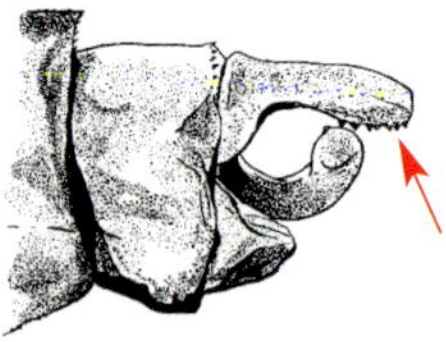

Note blunt tip of male cerci in side view.

Description: Adults average 2.0 inches long.

A beautifully contrasting, dark and light species. Male's thorax is green with 1st and 2nd dark shoulder stripes in sinuous curve (hence common name) and only slightly separated. Abdomen is mostly yellow with dark triangles on sides enlarged to form apparent bands at ends of most segments. Abdomen tip slightly clubbed with wide, yellow margins. Eyes blue-gray; face pale and unstriped. Female similar to male in color, but with a stouter, less clubbed abdomen. She has two pairs of small horns at rear of head (occipital and postoccipital horns).

Identification Clues: Curved, slightly separated 1st and 2nd dark shoulder stripes (↑) on green thorax (↑) separate this species from other snaketails. Further differs from Pale Snaketail in having greener thorax, more black on abdomen and differently shaped male

Female shows more yellow and light brown rather than green on her abdomen.

claspers with tips of cerci blunt (↑) (pointed on Pale—see also pg. 177). Separate snaketails in hand from other clubtails by their three-celled anal loop (pg. 176).

Distribution: Uncommon in northern part of Northern Rockies in Foothills and lower Montane Zone.

Life Cycle & Behavior: Adults have an unusually long flight period for a snaketail, emerging in June and flying through September. Males make long patrols and may fly far from water. Often perches on low shrubs, sometimes rocks or ground. Males are most often seen in afternoon. Female lays eggs unattended by making swift, low flights and tapping the water surface with her abdomen to wash off eggs. Occasionally she will splash down, for unknown reasons. Nymphs likely take three years to develop and sometimes emerge in huge numbers not far from water's edge.

Spiketails Family Cordulegastridae

Members of this stunning family are characterized by their large size, contrasting black and yellow body colors with blue or green eyes and their penchant for small forest streams and seeps. Defining features of the genus that set them apart from others are the eyes which just meet at one point on top of the head, and the females' spike-like ovipositor (actually an enlarged subgenital plate). During mating this ovipositor fits into a depression on the underside of the male's thorax and base of his abdomen. The abdomen is long and slender, slightly enlarged at the tip of males, but not females. Spiketails differ from the cruisers—the only other family with which they are apt to be confused—by having two yellow stripes on the side of the thorax instead of one, much smaller eyes and legs not nearly as long. They fly low in sunny clearings on brief feeding forays. Males routinely fly long beats up and down streams within a foot of water. Females lay eggs by hovering over mud or very shallow water and plunging their ovipositors into the bottom. The nymphs are large and very hairy with jagged teeth on the median edges of the "mask" that covers their face. Worldwide there are 51 species in this family, with nine species in North America, all in the genus *Cordulegaster*. Here in the Rockies we have just one species, the Pacific Spiketail, but it's a spectacular one!

Spiketails (Genus *Cordulegaster*)

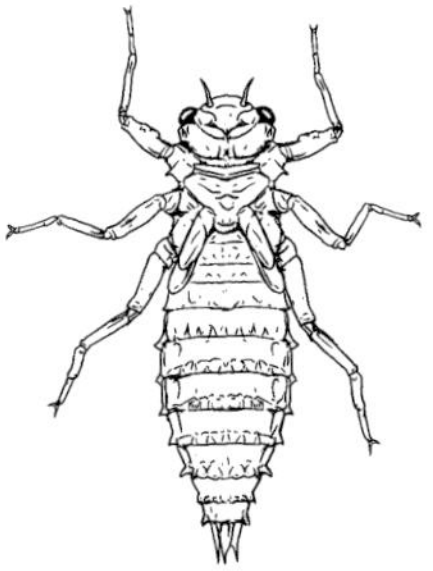

Typical nymph of the family Cordulegastridae; They are large, hairy burrowers. They lay in wait until suitable prey happens by to trigger the awesome, toothed, grasping lip.

Cruisers
Family Macromiidae

This is a fascinating family with a number of unusual characteristics. Defining features include a single, diagonal pale stripe on the side of the thorax, extremely long legs, large eyes in contact on top of the head and large overall size. It is primarily a group of rivers and streams, but most species are also at home along the shores of large lakes. The abdomen is usually slightly clubbed at the tip (males only). The single pale side stripe on the thorax easily separates this family from all other large dragonflies. Males cruise easily with their large wings for long distances up and down rivers and streams, and along lakeshores. They perch obliquely or vertically on vegetation, often hanging similar to the darners. Females oviposit by flying fast and low, tapping their abdomen here and there to wash off eggs. The nymphs are large, sprawling creatures with short, rounded abdomens. They are easily recognized by a prominent, triangular, up-curving horn on the head before the eyes and by their extremely long legs that give them a spider-like look. The nymphs are renowned for often crawling long distances to emerge, up to 80 feet from water. This family has at various times been included as a subfamily within the emeralds (family Corduliidae) and the skimmers (Family Libellulidae); the taxonomy of the group remains controversial. Their many unique characteristics seem to merit their retention as a family, pending further developments. Worldwide there are 121 species in this family, with nine species in two genera in North America. Here in the Rockies we have just one species in the genus *Macromia* (River Cruisers), and it's an impressive one, the Western River Cruiser.

River Cruisers (Genus *Macromia*)

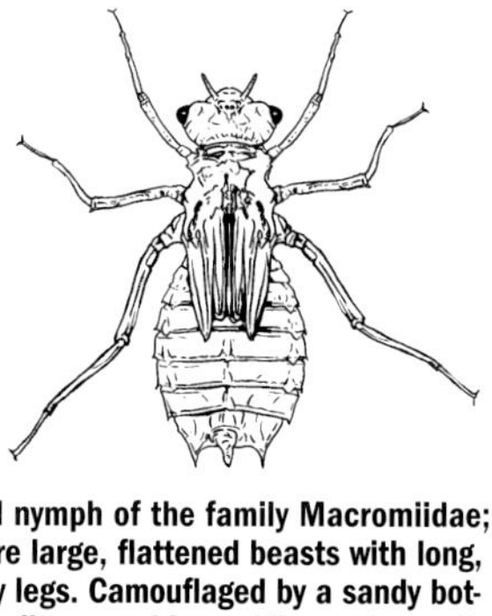

Typical nymph of the family Macromiidae; They are large, flattened beasts with long, spidery legs. Camouflaged by a sandy bottom or silt-covered in muddier environs, the nymphs lie in wait for their food to come drifting past.

Pacific Spiketail *Cordulegaster dorsalis*

APRIL	MAY	JUNE	JULY	AUG	SEPT	OCT

Breeding habitat: Small, clear streams ranging from mountain torrents to drainage ditches in forested landscapes.

Nature Notes:

One of the largest dragonflies in the Rockies and probably the most spectacular in appearance.

Females are rarely seen except when ovipositing.

Nymphs have been observed crawling one to five feet up the trunks of streamside alder trees to emerge.

Scientific name *Cordulegaster* alludes to the clubbed abdomen tip of males in this genus. Species name *dorsalis* means "dorsal," probably referring to the conspicuous, yellow abdominal spots.

Description: Adults average 3.1 inches long.

A magnificent insect in both size and coloration. The body is blackish with contrasting yellow stripes on the thorax; a pair on top and two wide pairs on each side. The abdomen prominently spotted (↑) with yellow on segments 2 to 9, spots slightly divided along the midline. Eyes blue, forehead yellow. Female colored like male but abdomen stouter and with long, pointed ovipositor that extends well past abdomen tip (↑).

Identification Clues: Only spiketail likely to be encountered in Rockies. Large size, black and yellow coloration and eyes barely touching at a point (↑), easily separate it from other families. Note that Western River Cruiser has only a single pale stripe on each side of the thorax, not two as Pacific Spiketail has.

Female spiketails have greatly enlarged subgenital plates used to deposit eggs into stream bottoms. Although not true ovipositors, thay are often refered to as such.

Distribution: Foothills and Montane Zone of the western half of the Northern Rockies and scattered pockets in the Central and Southern Rockies.

Life Cycle & Behavior: Adults emerge primarily in June and fly into September. Males patrol large reaches of streams with low, deliberate flights. Often forages in sunny clearings and ravines. Females lay eggs near stream edge while hovering by plunging abdomen tip with repeated thrusts into sand or silt under an inch or so of water. Nymphs require three or four years to develop and spend final winter in their last instar.

Western River Cruiser *Macromia magnifica*

APRIL	MAY	JUNE	JULY	AUG	SEPT	OCT

Breeding habitat: Clean, mid-sized streams to large rivers, often with sand bottoms; also large, open lakeshores.

Nature Notes:

Flies most dependably at water in the late morning during calm weather.

To observe them, pick a good vantage point along the stream and wait. No use in chasing them because of their blazing speed.

Scientific species name *magnifica* means "magnificent," probably alluding to their large size and the riveting appearance of the abdomen.

Description: Adults average 2.8 inches long.

A large dragonfly, remarkable for its appearance and behavior, as well as its size. The male's thorax is brownish gray with a pair of short, pale yellow stripes on top and a single long, oblique pale stripe on each side (↑). The abdomen is black with large, off-white spots on top and at the bases of segments 2 to 8, giving a banded look. Eyes bluish gray; face white with a black cross line. Female colored like male, but abdomen stouter and pale top spots may be larger.

Identification Clues: Unmistakable. Only cruiser likely to be found in the Rockies. Cruisers easily distinguished from all other families by single pale side stripe on thorax. Note also long legs, large overall body size, absence of blue on body and large eyes.

Note the long legs and single prominent side stripe on the thorax of this female.

Distribution: Uncommon in the Foothills, perhaps into lower Montane Zone, in the south-west portion of the Central Rockies, including the Uintas.

Life Cycle & Behavior: Adults emerge primarily in June and fly into September. Males patrol large reaches of streams and rivers early in the day with swift but bouncy flight. Forages late in the day in open fields. Mating is lengthy and occurs in trees. Female lays eggs by flying fast and low, tapping her abdomen every one or two yards, releasing eggs; she may also hover and tap vigorously. Little is know about the life history of the nymphs, but they are known to crawl up to 45 feet overland to emerge.

Emeralds
Family Corduliidae

The emeralds are a wonderfully interesting family of small to mid-sized odonates, even though they are among the least colorful. Most members are simply a shiny dark brown or black with small pale markings and many show a green or bronze metallic sheen on the thorax. They get their common name from the bright green, jewel-like eyes of most species when mature (reddish brown when immature). Many species are uncommon and elusive, and their breeding habitats can be puzzlingly hard to find, which naturally adds to their allure. Structurally they are most similar to the skimmers (Family Libellulidae) and some authors combine the two into one family. I prefer to retain them as a separate family. Differences between the two families are given in the family key (pg. 45).

The emeralds are a primarily northern family, well adapted to cold climates. They inhabit a wide range of habitats including lakes, ponds, bogs, fens and streams. Many are quick and tireless fliers, hovering occasionally and spending much of the day on the wing. Most perch obliquely or hang vertically when resting. A readily seen characteristic in the hand is the vaguely foot-shaped anal loop of the hind wing, with a rounded, but not pointed, "toe" (illustration). Females have subgenital plates of various lengths, modified on some species into a conspicuous spout-like structure (often called a pseudo-ovipositor). Worldwide there are 240 named species, with 50 species in North America in seven genera. In the Rockies we have eight species in three genera.

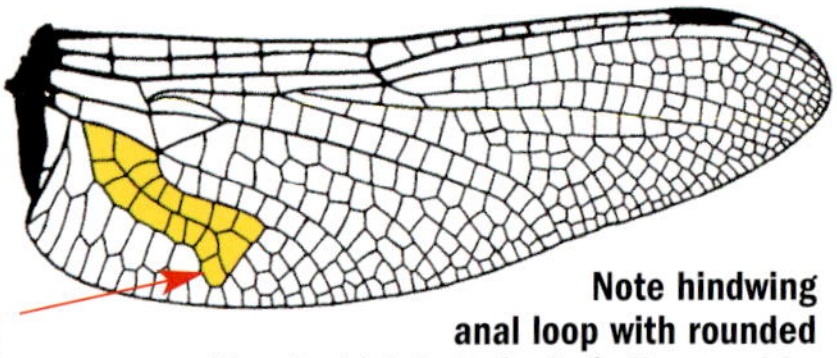

Note hindwing anal loop with rounded "toe," which is typical of all emeralds. Contrast with skimmer anal loop, pg. 214.

The Common Emeralds — Genus *Cordulia*

There is just one North American species in this tiny genus, the American Emerald, and we've got it in the Rockies. The genus is most similar to the striped emeralds, but it is easily distinguished from them and from other brown dragonflies by the uniquely forked epiproct of the males. There are slight differences in wing venation as well, the most important of which for our identification purposes is the presences of only one cross vein in what is known as the cubito-anal area of the hind wing (illustration on page 195); striped emerald species have two crossveins there. The genus name *Cordulia* means "clubbed," referring to the slightly clubbed abdomen tips of the males.

The Striped Emeralds — Genus *Somatochlora*

Yippee! Now I get to tell you about my favorite odonate genus! So, what's the big deal with the striped emeralds? Well, let's start with a few characteristics of the group. With their slender, shiny, mahogany brown to black bodies, metallic green and bronze reflections on the thorax and brilliant green eyes, they are the picture of understated beauty and elegance. They are graceful, artful fliers and some species possess blazing speed. They have been called the "Ferrari's" of the odonate world by some. Their habitats, though diverse, tend to be wild, challenging to access and often remote. Some species are shy and elusive, and populations are rarely abundant, even in the heart of their range and in prime habitat. The great Canadian odonatist E. M. Walker referred to them as the "shy denizens of the wilderness." A number of species are quite rare, or their populations are highly localized. In some cases their habitat, the life blood of any species, is at risk because of development or other factors. In fact, the only federally endangered odonate in North America is a striped emerald, the Hines Emerald. The life histories of some odonate groups are poorly known because few have bothered to study them. The life histories of many *Somatochlora* species are poorly known, not because of a lack of interest, but because their habitats are remote, or because specialists have had much trouble trying to find their breeding sites with any consistency. To heap challenge upon challenge, even when you do find a breeding site, the nymphs are often thinly distributed, hidden in difficult to reach spots and nearly impossible to identify when you're lucky enough to catch one. Even the adults can be difficult to identify because of similarities among groups of species. So there you have it—elegance, understated beauty, aerial mastery, blazing speed, elusiveness, rarity, found in remote and often pristine habitats, difficulties in identification—in other words, an unending stream of compelling attributes and challenges for those who would study them. All of this has contributed to an allure that has led to a near cult-like following among some odonate enthusiasts.

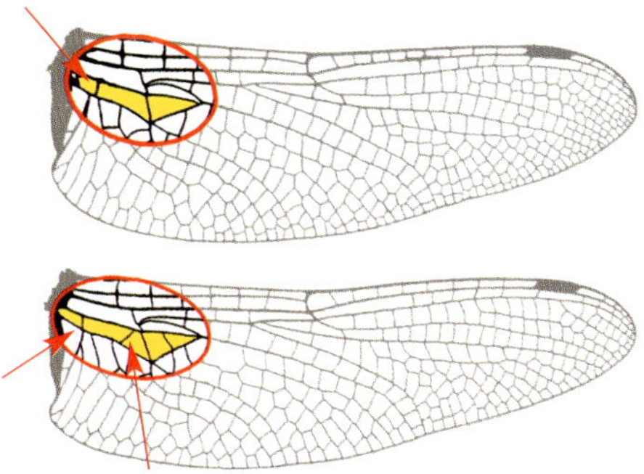

Emerald hindwings: Top illustration is American Emerald showing one cross vein in cubito-anal area. Bottom illustration is striped emerald showing two cross veins there.

This is the most diverse genus of North American emeralds, with a generally northern distribution, often boreal, and sometimes in tundra. They are typically found in bogs and fens, sometimes lakes, and there are some stream species. Many species are conspicuously marked with yellow spots or streaks on the sides of the thorax that tend to become duller with age. Some species have pale rings (called annuli) on at least the middle abdominal segments. Males have distinctively shaped cerci, and the subgenital plates of females come in a wide variety of sizes and shapes, from short and wide to spout-like "pseudo-ovipositors." Therefore, individuals should be netted when possible and those structures examined carefully with a hand lens until familiarity with each species is achieved. Even with familiarity, some species can only be reliably identified in the hand. The nymphs are crawlers on mosses and other vegetation in their habitats. Nymphs of stream species are often found tucked away deep beneath undercut banks. Most species take at least two years to develop. The genus name *Somatochlora* means "green bodied" referring to the extensive metallic green reflections on the thoraxes of many species.

The Baskettails — Genus *Epitheca*

This is a genus of mid-sized, rather dull-colored odonates with hairy thoraxes that largely lack the metallic luster of other emeralds. Even the green of their eyes is rather dull. But they don't disappoint. They are remarkable aerialists, quick and sure in their movements, and spend many hours on the wing, often in huge aggregations. These swarming groups must consume incalculable numbers of tiny insect prey. For me, seeing swarming hordes of baskettails in the spring exemplifies the fullness of that season with all its energy and vitality. Females have uniquely long, bilobed subgenital plates that they use to lay eggs like no other group. Sid Dunkle (2000) has described the process that gives the genus its name more vividly than most: "Females use their subgenital plate like a forklift to carry their yellow to orange eggs, resembling a miniature sea plane with their cargo of eggs 'in one basket.'" Females deposit this load of "cargo" on a nearly submerged aquatic plant where it unravels into a single string of several hundred eggs up to six-inches long. North American species in this genus are sometimes split into two genera by some authors. In this view, our single species, the Spiny Baskettail, would fall into the genus *Tetragoneuria.*

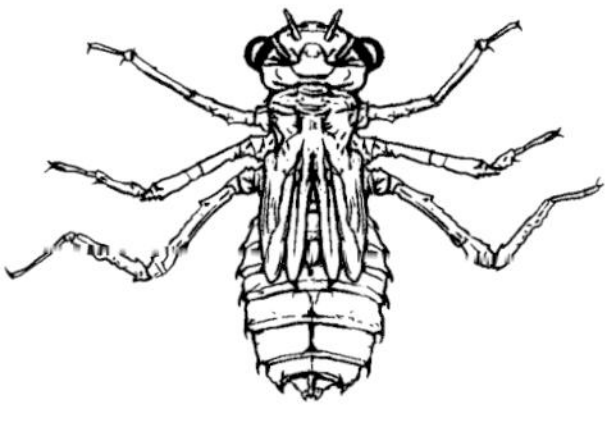

Typical nymph of the family Corduliidae; Some are sprawlers and some are crawlers. The sprawlers spend their time awaiting prey camouflaged amongst the bottom detritus. Crawlers pursue prey in the greenery of their aquatic world.

"Striped Emerald" *Somatochlora* Appendages Comparison

Grouped together as the "striped emeralds," the *Somatochlora* are best differentiated by their diagnostic terminal appendages. Male claspers are distinctive, as are female subgenital plates in most species. In-the-hand is the best way to observe these traits but a close-up photo of the abdomen tip may be enough to make an I.D.

Male lateral | Male dorsal | Female lateral

Ringed Emerald (pg. 200) *Somatochlora albicincta*

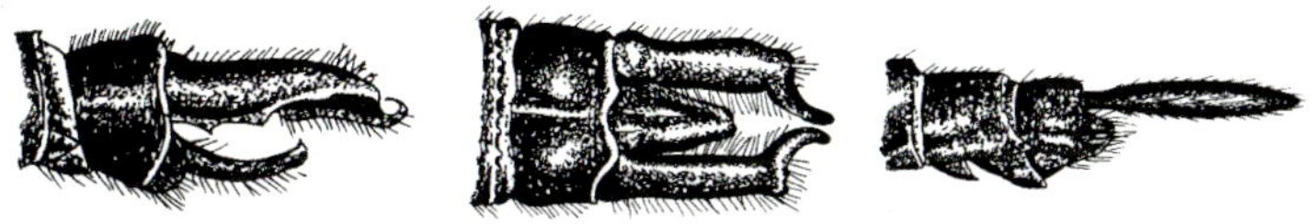

Hudsonian Emerald (pg. 202) *Somatochlora hudsonica*

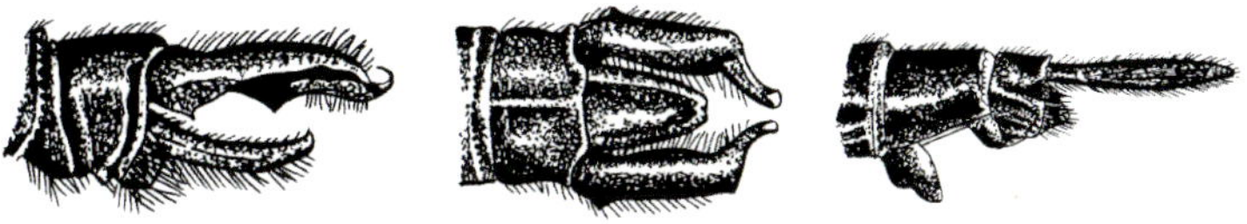

Mountain Emerald (pg. 204) *Somatochlora semicircularis*

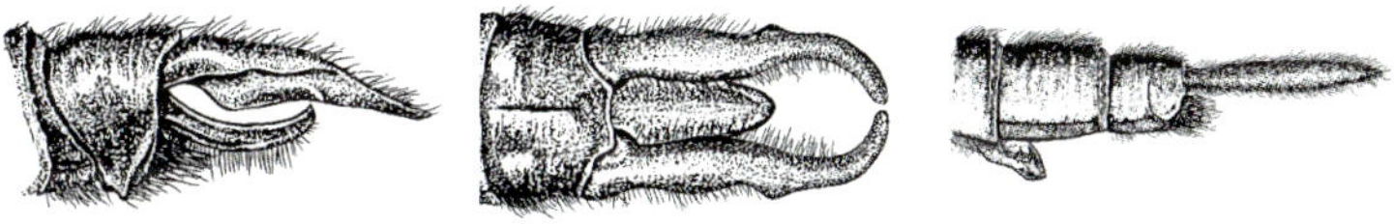

Ocellated Emerald (pg. 206) *Somatochlora minor*

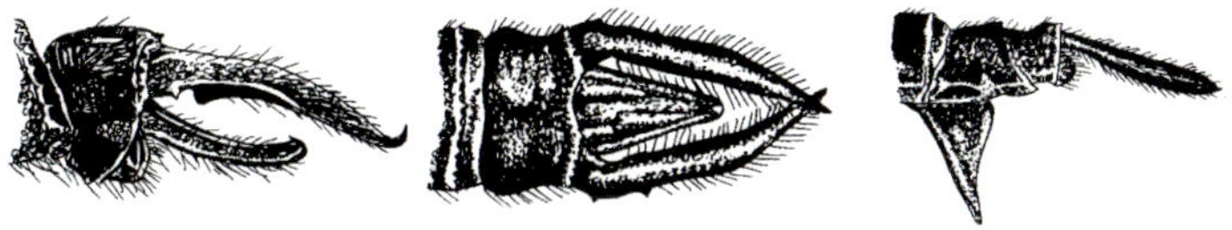

Brush-tipped Emerald (pg. 208) *Somatochlora walshii*

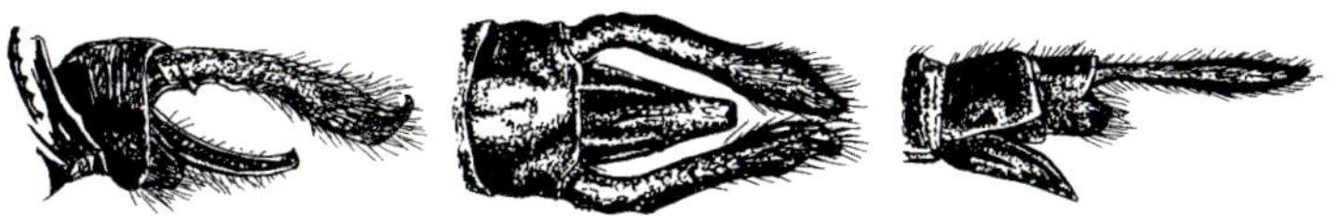

Plains Emerald (pg. 210) *Somatochlora ensigera*

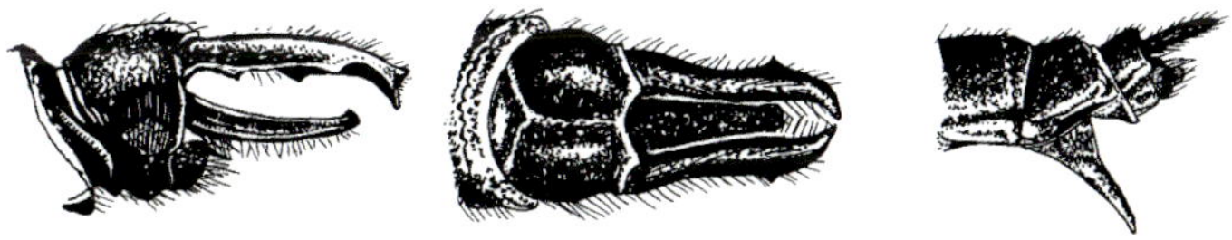

American Emerald *Cordulia shurtleffii*

APRIL	MAY	JUNE	JULY	AUG	SEPT	OCT

Breeding habitat: Variety of ponds—especially sphagnum bog ponds—and small lakes, usually in forested landscapes.

Nature Notes:

Latin name often spelled *C. shurtleffi*.

Males patrol shifting territories, along one area of shoreline for a few minutes, then moving elsewhere.

A population found in a temporary pond at 10,400 feet in Gunnison County, Colorado, by Ruth Willey in 1969-70, was still there when I visited the site in 2007

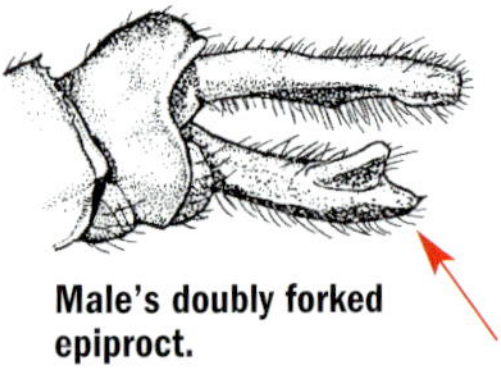

Male's doubly forked epiproct.

Description: Adults average 1.9 inches long.

Male's thorax is brown and hairy with metallic green highlights around sutures. Abdomen is dark brown to blackish and slightly clubbed at the tip. A conspicuous but narrow pale ring is between segments 2 and 3 and the sides of segments 1 and 2 have brownish orange markings. Male epiproct deeply forked and each tip also notched (↑). Eyes brilliant green, forehead metallic green, wings clear. Female similarly colored with stouter abdomen lacking club and wings gaining brownish tint with age. Her subgenital plate a quarter the length of segment 9 and split halfway to the base forming two lobes (↑).

Identification Clues: Most similar to striped emeralds, but most of those have either yellow marks on sides of thorax or white rings around most abdominal segments, both lacking on American. Also compare cubito-anal cross

Female's subgenital plate does not project downward as it does with most striped emeralds. Note male's slightly clubbed abdomen and pale ring between segments 2 and 3 (inset).

veins between the genera (pg. 195). Mountain Emerald is slightly larger and abdomen more slender. Also, white ring dividing segments 2 and 3 more conspicuous than on some striped emeralds, though all have the ring. Doubly forked epiproct (lower clasper) of male (↑) and bilobed subgenital plate of female (↑) both diagnostic in hand.

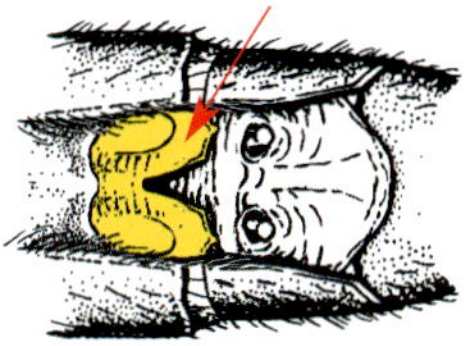

Bilobed subgenital plate of female.

Distribution: All Rocky Mountain regions in Foothills through Subalpine Zone. Absent in southern half of Southern Rockies.

Life Cycle & Behavior: Adults emerge early, in May and June and fly into August. Males patrol pond edges at waist height darting quickly and hovering often. Often seen hunting in forest glades, sometimes in swarms mixed with other emeralds. Mating is lengthy and occurs at least in part while flying. Female oviposits while in swift, low flight, striking the water surface with her abdomen periodically, releasing eggs. Life history of the nymphs not yet studied in the mountains. Likely overwinters in final instar.

Ringed Emerald *Somatochlora albicincta*

APRIL	MAY	JUNE	JULY	AUG	SEPT	OCT

Breeding habitat: Ponds, lakes and slow streams, often boggy and with emergent plants along shore.

Nature Notes:

Populations vary in body size and coloration across its range, and subspecies have been named.

May be more active before noon, and therefore more easily observed then.

Noted for its aggressive behavior, sometimes chasing other dragonflies larger than itself.

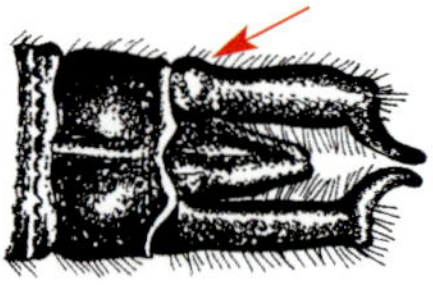

Male's abdomen tip in dorsal view. Angle at base of cerci less prominent than Hudsonian (pg. 202).

Description: Adults average 1.9 inches long.

Male's thorax is bronze with metallic green reflections and a short, diamond-shaped, pale side stripe. Abdomen is shiny black with thin white rings around segments 2 to 9 (↑) (*albicincta* means "white ringed"). Cerci in top view are straight for most of their length then sharply angled inward near tips; in side view, tip is upcurved (pg. 197). Eyes brilliant green; face black with yellow on sides before eyes. Female very similar to male, with the subgenital plate short, about $^{1}/_{3}$rd as long as segment 9 and notched (pg. 197 and see Hudsonian, pg. 203).

Identification Clues: Ringed and Hudsonian are only two white-ringed emeralds reliably reported from Rockies and they often fly together—not easily confused with other groups. Separate slightly smaller Ringed male

Striking green eyes and prominent white abdominal rings on this female produce a visually pleasing image.

from Hudsonian in the hand by less prominent lateral angle near base of cercus (↑), best seen in dorsal view. Female Ringed's short, notched subgenital plate contrasts with the longer, rounded, projecting subgenital plate of Hudsonian. Also, Ringed often has two tiny pale spots on segment 10 near base of cerci, lacking on Hudsonian. Location a clue; Ringed is found only in Northern Rockies.

Distribution: Uncommon in northern half of the Northern Rockies in Montane Zone.

Life Cycle & Behavior: Adults emerge in June and July and fly into September. Males patrol water edges slowly at knee height or lower, hovering occasionally. Forages in openings near breeding areas and not very high (within reach of net). Female oviposits while in slow flight, rhythmically tapping the water surface with her abdomen tip to release eggs. Nymph likely has a two year life cycle, overwintering in a late instar.

Hudsonian Emerald *Somatochlora hudsonica*

APRIL	MAY	JUNE	JULY	AUG	SEPT	OCT

Breeding habitat: Ponds, lakes and slow streams, usually boggy; muskeg pools with firm peaty edges.

Nature Notes:

Occurs at higher elevations than similar-looking Ringed Emerald.

Often occurs with Ringed Emerald and hybridizes with that species in northern Yukon, but not known to do so in U.S. Rockies.

Found at 10,200 feet in the Uinta Mountains.

Male's abdomen tip in dorsal view. Angle at base of cerci more prominent than Ringed (pg. 200)

Description: Adults average 2.1 inches long.

Male's thorax is bronze with metallic green reflections and a short, thin, dull yellow side stripe. Abdomen is shiny black with thin white rings around segments 2 to 9 (↑). Cerci in top view are straight for most of their length then sharply angled inward near tips; in side view, tip is upcurved (pg. 197). Eyes brilliant green; face black with yellow on sides before eyes. Female very similar to male with a rounded, projecting subgenital plate nearly as long as segment 9.

Identification Clues: Hudsonian and Ringed are only two white-ringed emeralds reliably reported from Rockies and often fly together—not easily confused with other groups. Separate slightly larger Hudsonian male from Ringed in the hand by larger lateral angle (tooth) near base of cercus (↑), best seen

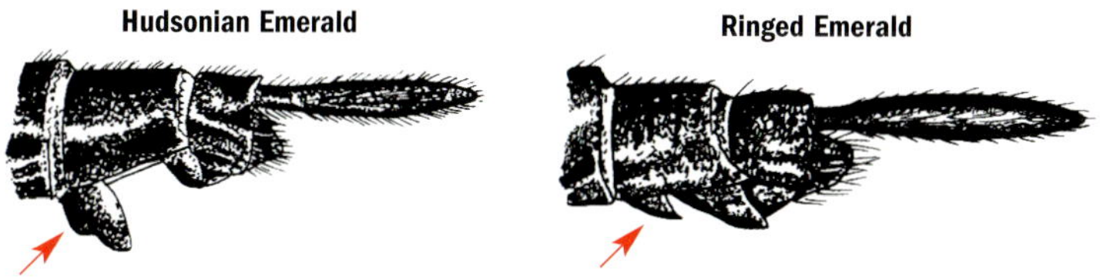

Hudsonian Emerald female (left) has a longer subgenital plate than Ringed Emerald (right).

in dorsal view. Female Hudsonian's rounded, projecting subgenital plate contrasts with short, notched subgenital plate of Ringed (↑). Also, Hudsonian lacks pale spots on segment 10 near base of cerci, often present (but tiny) on Ringed. Location a clue; Hudsonian more widespread, Ringed occurs only in Northern Rockies.

Distribution: Parts of all Rocky Mountain regions in Montane and Subalpine zones.

Life Cycle & Behavior: Adults emerge in June and early July and fly into September. Males patrol water edges slowly at knee height or lower, looking for females. Foraging and nearshore oviposition similar to that described for Ringed Emerald. Nymph likely has a two year life cycle, overwintering in a late instar.

Mountain Emerald *Somatochlora semicircularis*

APRIL	MAY	JUNE	JULY	AUG	SEPT	OCT

Breeding habitat: Ponds, sometimes temporary

Nature Notes:

Common and widespread throughout the Rockies, and no stranger to heights above 11,000 feet in Colorado.

Nymphs at high-elevation temporary ponds take four years to mature and survive periods of drying by hiding under logs, rocks and moss mats or burrowing into bases of sedge clumps. They lose body water at a slower rate than most odonates so can survive desiccation longer.

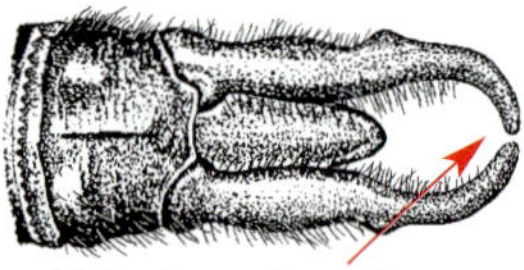

Distinctive male cerci

Description: Adults average 2.0 inches long.

Male's thorax is metallic green with bronze reflections and a pair of oval yellow spots on each side that become obscured with age. Abdomen black and unmarked except slightly with yellow near base. Cerci in dorsal view nearly straight for most of their length then roundly incurved toward tip forming a half circle (↑) (hence scientific name). Eyes brilliant green; face black with yellow on sides before eyes. Female similar to male, but with paired orange-brown spots on top/side of segment 3. Her subgenital plate rounded and projected obliquely (↑), shorter than segment 9.

Identification Clues: Lacks white rings on abdomen possessed by Ringed and Hudsonian Emeralds. Abdomen longer and more slender than Ocellated or Brush-tipped Emeralds. Thorax side spots not bright yellow like Plains.

Note that females's subgenital plate is short and obliquely projecting.

Male cerci diagnostic, much different than any of those (pg. 197), so netting is helpful. Females tougher; compare subgenital plates with other emeralds (pg. 197). See American Emerald pages (pg. 198-199) for characters to separate it.

Distribution: All Rocky Mountain regions in Montane and Subalpine zones.

Life Cycle & Behavior: Adults emerge in June and July and fly into September. Males patrol small territories along pond edges or low over grasses and sedges in wet meadows, hovering often. Often seen hunting in sunny clearings from near ground to high in trees. Male grabs female while she rests, then pair goes on nuptial flight while copulating, land in shrubs or trees after few minutes and continuing to copulate for up to 25 minutes. Female oviposits while in slow, low flight, stopping to tap water surface with her abdomen tip, then moving to another spot. Nymph stage has been well studied by Ruth Willey in temporary ponds at 10,400 feet in Colorado (see Nature Notes).

Ocellated Emerald *Somatochlora minor*

APRIL	MAY	JUNE	JULY	AUG	SEPT	OCT

Breeding habitat: Clear, small to mid-sized streams with slow to moderate flows in or near forested landscapes.

Nature Notes:

Male patrols very close to water near breeding site, but forages at head height or higher in forest glades.

Often perches on grasses and sedges after lengthy periods of flight.

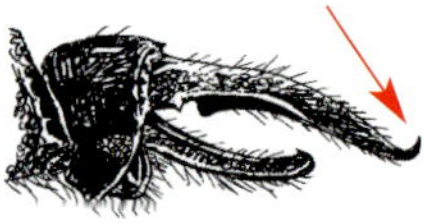

Male claspers in side view, Note upturned tip of cerci.

Description: Adults average 1.8 inches long.

Male's thorax brown with metallic green reflections and a pair of small, bright yellow oval spots on each side (↑). Abdomen is dark, spindle-shaped and short, about as long as hind wing, with yellow spots near base on segments 2 and 3. Cerci in dorsal view slender and gradually angled in toward tips (may be crossed). Tips of cerci in side view upturned (↑) (ski-type—pg. 197). Eyes green; face metallic black/green with dull yellow marks in middle and on sides. Female like male, but her wings often amber-tinted. Her subgenital plate yellowish and spout-shaped and perpendicular to the abdomen (↑) and cerci long (pg. 197).

Identification Clues: Combination of small size, short abdomen and small, nearly round yellow spots on side of thorax separates this species from other emeralds. In the hand, note

Round thorax side spots and prominent spout-shaped subgenital plate readily identify this female.

male's upturned tips of cerci are not hairy, and female's spout-like subgenital plate, though large, is no longer than her cerci. Most similar to Brush-tipped Emerald, but above the suite of characters and lack of mid-abdominal spots separates them. Plains Emerald is somewhat similar but larger and with different thoracic yellow marks and dissimilar cerci (male's cerci tips not upturned and female's cerci very short).

Distribution: Foothills and Montane Zone in much of Northern Rockies and limited parts of Central and Southern Rockies.

Life Cycle & Behavior: Adults emerge in June and fly into September. Tenerals disperse far from breeding sites to feed and mature. Males patrol quickly up and down streams, stopping often for lengthy hovering. Often forage in sheltered, sunlit forest openings. Female oviposits alone and secretly, flying close to the bank and low over water, stopping to tap her abdomen tip in water and sometimes a few inches above water on mossy, muddy banks. Nymphs develop in two years and spend last winter in final instar.

Brush-tipped Emerald *Somatochlora walshii*

APRIL	MAY	JUNE	JULY	AUG	SEPT	OCT

Breeding habitat: Small, shallow streams with barely perceptible flow through bogs, fens, cedar swamps and marshes.

Nature Notes:

Often found coursing at waist height along small roads, especially unimproved ones.

Common name alludes to male's heavily bewhiskered cerci.

Most I ever saw in one place were scattered in a huge mixed swarm of basketails, outnumbered about ten to one.

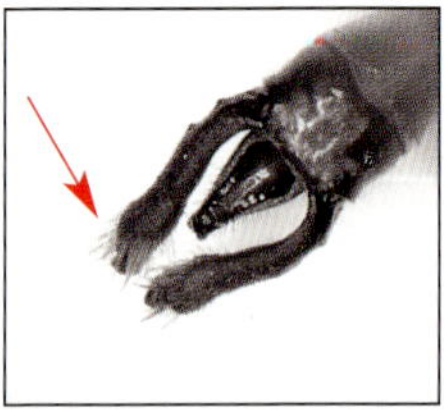

The males "brush-tipped" cerci from above.

Description: Adults average 1.9 inches long.

Male's thorax metallic bronzy green with pair of bright yellow marks on each side, the first (anterior) a thin stripe, the second (posterior) shorter and oval. Abdomen is dark, spindle-shaped and short, about as long as hind wing, with yellow spots near base and dull yellow lateral spots on at least segments 5 to 7. Cerci in top view slightly bowed at base and very brush-like with dense hairs at tips (↑). Tips of cerci in side view upturned (ski-type) but largely obscured with hairs (pg. 197). Eyes green; face metallic black/green with dull yellow marks on sides. Female like male, but her wings often orange or amber-tinted, especially when teneral. Her subgenital plate dark and scoop-shaped and obliquely projecting from abdomen (↑) (pg. 197); cerci long (↑).

Note the long cerci and obliquely projecting subgenital plate on this female.

Identification Clues: In the hand, male's upturned tips of cerci partially obscured by dense brush of hairs is diagnostic (pg. 197). Size small and abdomen short for a striped emerald, plus slender yellow front stripe and oval spot on side of thorax are excellent field clues. Female's size, shape, striping and obliquely projecting subgenital plate (pg. 197) also diagnostic. See Ocellated and Plains Emerald pages for additional clues.

Distribution: Uncommon in northern tip of Northern Rockies in Foothills and Montane Zone.

Life Cycle & Behavior: Adults emerge in June and early July and fly through August. Males patrol along slow streams at waist height darting quickly and hovering often. Often seen hunting in forest glades, sometimes until dusk in swarms mixed with other emeralds. Female stops slow flight to oviposit by tapping repeatedly in small openings and around reed clumps, then moving to new site. Nymphs likely develop in two years and spend last winter in one of last two instars.

Plains Emerald *Somatochlora ensigera*

APRIL	MAY	JUNE	JULY	AUG	SEPT	OCT

Breeding habitat: Streams and ditches in open or partially forested areas.

Nature Notes:

Males are territorial, chasing each other from preferred stream sections.

Ken Tennessen and I found nymphs deep beneath undercut, grassy banks, only in deeper areas and near pool constrictions in a North Dakota stream.

Description: Adults average 2.0 inches long.

Male's thorax dark brown with metallic green reflections and a pair of bright yellow stripes on each side; first stripe long, second stripe ovate. Abdomen dark and longer than length of hind wing, with yellow marking only near base; yellow spot on lower side of segment 2 large. Cerci in side view squared off abruptly at tip with small, down-pointing spine (↑). Eyes green; face mostly yellow (↑), black on top of frons. Wings becoming brown-tinted with age. Female like male; her subgenital plate long, spout-shaped and perpendicular to her body; much longer than her very short cerci (↑).

Identification Clues: Other striped emeralds with bright yellow thoracic side stripes have mostly black faces, not mostly pale like Plains. Male's truncated cerci in side view are distinc-

Distinctive male cerci with down-pointing spine.

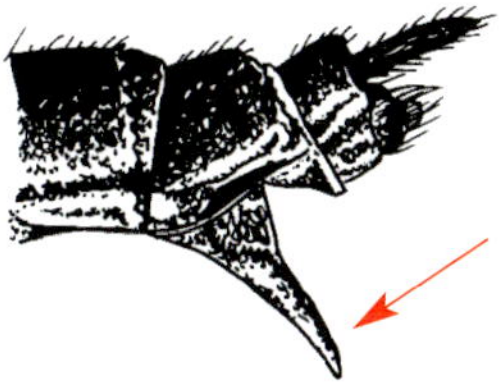

Female's spout-shaped subgenital plate.

tive (pg. 197) as are female's ultra-short cerci (pg. 197). Ocellated and Brush-tipped Emeralds both smaller.

Distribution: Prairie species rare in Foothills and Montane Zone of northern tip and eastern edge of Southern Rockies.

Life Cycle & Behavior: Adults emerge in June and July and may fly into September. Males patrol low beats up and down streams, hovering occasionally. Both sexes often perch on streamside grasses. Often seen foraging high in openings over or near streams. Female oviposits while in low flight, hovering to tap water with her abdomen tip under overhanging grasses; sometimes into mud or gravel above water line. Nymph stage is poorly known, especially near Rockies (see Nature Notes).

Spiny Baskettail *Epitheca spinigera*

APRIL	MAY	JUNE	JULY	AUG	SEPT	OCT

Breeding habitat: Ponds and lakes of various types, usually in forested landscapes; sometimes slow streams.

Nature Notes:

Egg clusters in water look like amphibian eggs.

Emerges synchronously in huge numbers from prime habitat in spring.

Nymphs are distinctive in appearance with lateral spines on segment 9 long and projecting rearward like "tail pipes."

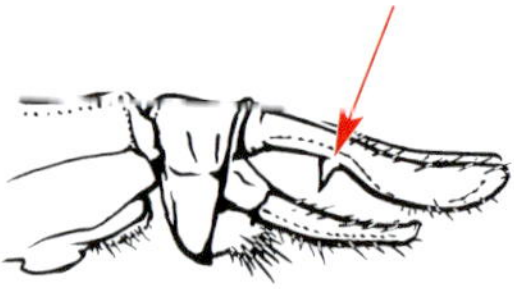

Male claspers in profile. Note the downpointing "spine."

Description: Adults average 1.9 inches long.

Male's thorax is two-tone brown and hairy. Abdomen is dark brown near base then black and spindle-shaped with yellow steaks on the sides of segments 3 to 8. Male cerci with downward pointing spine near midpoint (hence the common and scientific names). Eyes deep blue-green at maturity, brownish earlier; forehead with distinct, black T-spot; wings clear. Female similarly colored with greener eyes and cylindrical abdomen much stouter near base. Her subgenital plate with two long, incurved lobes separated near base.

Identification Clues: Overall appearance unlike any other genus and only one baskettail species likely in Rockies. Net some to examine male cerci in side view looking for downward pointing spine (↑)—if not there, may be

Note the deep blue-green eyes on this mature individual.

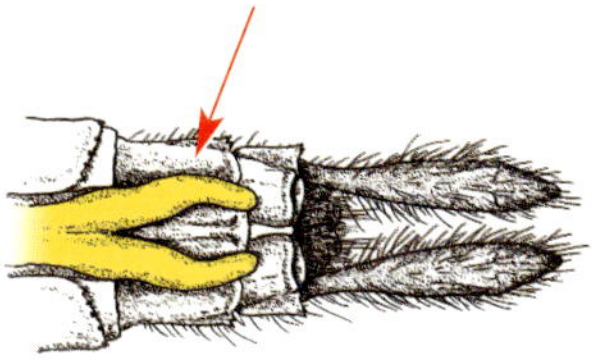

In ventral view female's subgenital plate is uniquely bilobed (highlighted).

another species of baskettail. Note also female's bilobed subgenital plate in ventral view (↑).

Distribution: Northern third of Northern Rockies in Foothills and Montane Zone.

Life Cycle & Behavior: Short, early flight period with adults emerging in late May and June and normally flying only into July. Males patrol low along pond edges and forest clearings near water until dusk, hovering often. Joins mixed-emerald feeding swarms in clearings, often far from water. Female oviposits while in flight, first with her abdomen tip canted upward, then tapping the water surface with it, releasing strings of eggs. Nymphs mature in one year and overwinter in final instar, ready to emerge as soon as spring temperatures are favorable.

Skimmers Family Libellulidae

The skimmers are a family full of mostly common species that range in size from small to moderately large. We have more skimmers here in the Rockies, both in numbers of species and numbers of individuals, than any other family of dragonflies. They are diverse in other ways as well, including colors, behaviors and life history strategies. Behavioral characteristics of the group include their showiness, tendency to perch often and horizontally or obliquely on vegetation around the margins of their breeding waters, and the ease with which many species can be approached and observed. Most species mate entirely in flight. Females lay eggs either in tandem (unusual among dragonflies) or with the male hover-guarding nearby. Physical characteristics useful in identification include: large eyes that touch on top of the head; body colors widely varied and unlike other groups (though rarely metallic); abdomens often short, stout and compressed; wings that are variously patterned with dark color in nearly half our species; and an anal loop in the hind wing that is distinctly "boot-shaped" with a pointed "toe." Females have subgenital plates that are usually short, and lacking a functional ovipositor. The hind wings of males are rounded at the bases, like females; in other dragonfly families, males have squared-off, angular lower bases of the hindwings. Pruinosity, especially of males, is much more common among skimmers than in other dragonfly families. Many aspects of wing venation are also distinctive in this family; when elements of wing venation are useful for separating species in the Rockies, they are described on the species pages. Structurally, skimmers are most similar to the emeralds (Family Corduliidae) and some authors combine the two into one family, but they are separately treated here. Differences between the two families are given in the family key (pg. 45).

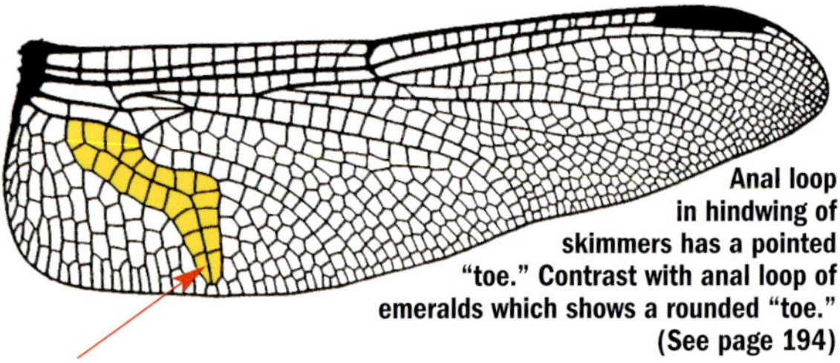

Anal loop in hindwing of skimmers has a pointed "toe." Contrast with anal loop of emeralds which shows a rounded "toe." (See page 194)

Identification of most skimmer species is fairly easy, which along with their ease of observation and commonness, means this is a great group for the beginner to get her feet wet with. Each genus tends to show a distinctive suite of characteristics. For example, most meadowhawks are small with predominantly reddish bodies, the whitefaces are also small with mostly black bodies and white faces, the king skimmers are large with variously patterned wings, the whitetails have wings heavily patterned with black and whitish-blue pruinescence on their abdomens that fairly glows, and so forth. So you should easily be able to place most skimmers you see into their proper genus by paging through the genera in this guide. Naturally, a few species are tricky, but would it be any fun otherwise? Once narrowed down to a genus, many species can be identified to species in the field using the identification clues on the species pages, but some genera, especially the meadowhawks and whitefaces, will have to be examined in hand until familiarity with them is attained.

The skimmers inhabit the full gamut of aquatic habitats, except for swift streams, but are most common around ponds, marshes and vegetated lakeshores. While not usually considered among the finest aerialists in the odonate world, the rainpool gliders and the saddlebags are adept at using the enormous surface areas of their hind wings to glide with apparent effortlessness with the winds for great distances. The Wandering Glider is well known to cross many hundreds of miles of ocean expanses in its travels. The nymphs are much like those of emeralds, either sprawlers that lie on silted bottoms or crawlers that climb about on aquatic vegetation. Many species have a one year life cycle, but some take two years or perhaps longer to develop, especially in the north or at higher elevations. Worldwide there are about 970 named species, with 108 species in North America in 25 genera. I treat *Plathemis* and *Ladona* as full genera; some authors include species in those groups in the genus *Libellula*. In the Rockies we have 31 species in nine genera.

The Whitetails — Genus *Plathemis*

Striking, unforgettable skimmers with wings heavily marked with black or dark brown (hereafter dark) and white or bluish-white pruinose abdomens on males that almost seem to glow, very conspicuous in flight or perched. This genus is included within the genus *Libellula* (king skimmers) by some authors. Whitetails are unusual in the great extent to which males and females differ in appearance. In fact, the female Common Whitetail looks much more like the female Twelve-spotted Skimmer (different genus) than it does to the male of its own species. There are only two species of whitetails in the world and we've got both of them in the Rockies!

The Corporals — Genus *Ladona*

The corporals are dark-bodied, habitual ground-perchers that get their name from a pair of wide pruinose bars on the top of their thorax that are reminiscent of the symbol for that military rank. They also have pruinescence on their abdomens, the extent of which varies among species. The nymphs are distinctive—a deep mahogany brown and slender in shape with long, sharply pointed anal appendages. This small genus with just three North

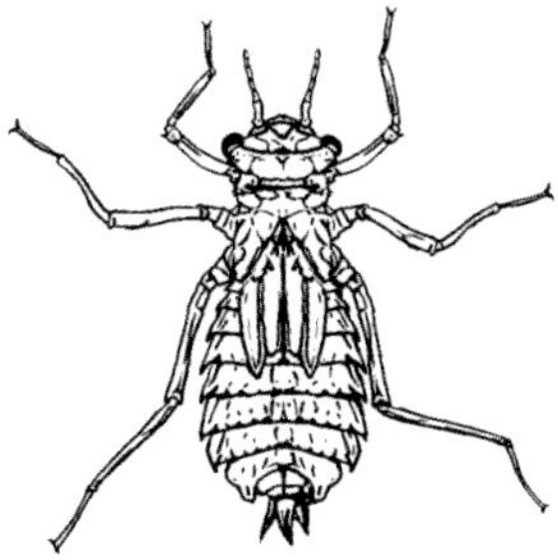

Typical nymph of the family Libellulidae; They are well-camouflaged crawlers, hunting amongst the green vegetation of still or slowly flowing waters.

American species has often been included within the genus *Libellula*, but since 1999 has been considered a valid genus by most authors. We have just one species in the Rockies, the Chalk-fronted Corporal.

The King Skimmers — Genus *Libellula*

The most easily recognized of our skimmers, large and showy, with dark markings on the wings of most species. Most species perch conspicuously on vegetation along side their breeding habitats. Mating is brief in all of our species, not much more than a minute and often much less, and usually occurs in flight. The abdomen is short, often visibly shorter than a hind wing. Many species develop pruinescence on the abdomen and thorax as they age. Females have lateral flanges on the sides of segment 8; these are extensions of the margins of the segment that are sometimes called "splash flaps." Females use them to splash eggs onto low banks while males typically hover-guard overhead. The nymphs are hairy sprawlers, often on dead or living plant material, and usually over soft muck or silt bottoms. We have seven species in the Rockies, all readily identifiable in the field by their wing markings, body colors and large size.

The Whitefaces — Genus *Leucorrhinia*

This is a circumboreal genus of small, northern, black-bodied dragonflies. They are an early emerging group with flight periods in spring and early summer. Defining physical features of the group are their bright white faces, small size, dark overall coloration, red or yellow markings on the abdomen and thorax, black legs and small, dark markings at the bases of the wings, which are larger and often triangular in shape on the hind wings. Members are easily distinguished from other genera in the field by the combination of the first four of the above characteristics. Males of some species develop whitish pruinosity at the base of the abdomen, though not usually the case in the Rockies, and aging females may become pruinose underneath the abdomen. The artfully designed combination of ivory white faces and bright red body marks on the jet black bodies of some species are quite beautiful. Characteristic behaviors include typically low flights close to shoreline vegetation, at still-water, often boggy habitats. Mating pairs usually perch on low vegetation near water. Males often hover-guard above and slightly to the rear of egg-laying females, and may hold competing males in tandem long enough for females to complete oviposition. This fascinating obstructionist strategy was coined "karate-guarding" by Sid Dunkle. Whitefaces are usually not difficult to approach, observe or net. Species identification can be tricky, especially of females and teneral individuals, until familiarity with subtle color pattern differences is attained. Fortunately, all species have distinctively shaped reproductive parts and wing characteristics that are readily seen with a hand lens, so frequent netting is a great way to learn this group. Nymphs crawl about on submerged plant material, develop in one year and overwinter in the final instar, thus being ready to emerge as soon as water temperatures become suitable in spring.

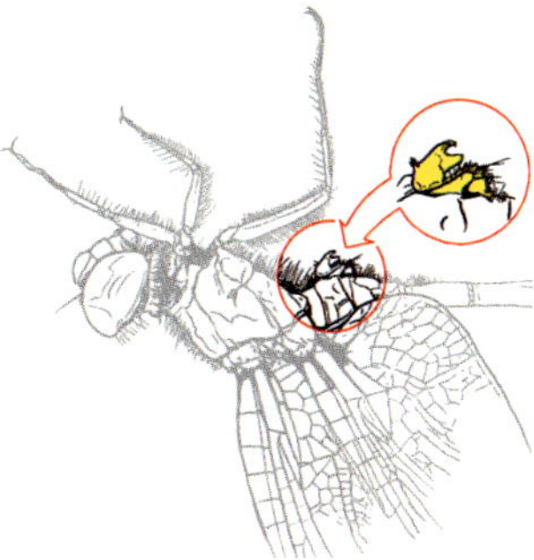

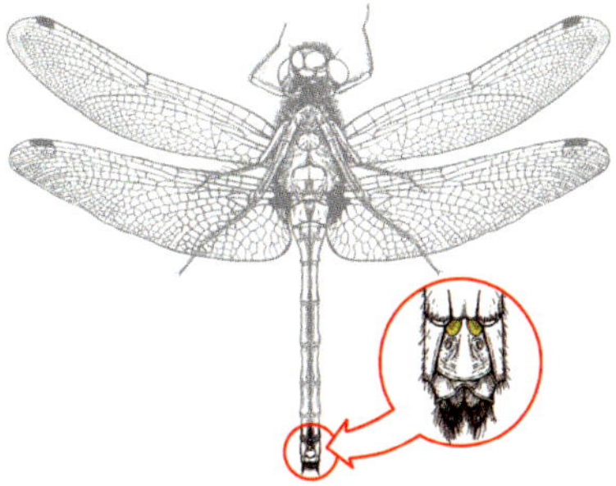

Male hamules (lateral view) | **Female subgenital plate (ventral view)**

Dot-tailed Whiteface (pg. 242) *Leucorrhinia intacta*

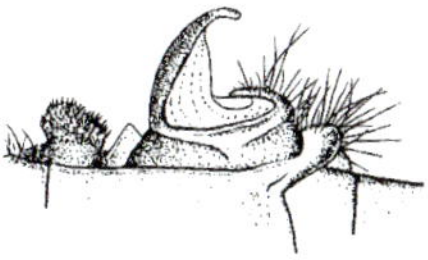

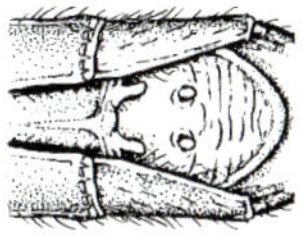

Hudsonian Whiteface (pg. 244) *Leucorrhinia hudsonica*

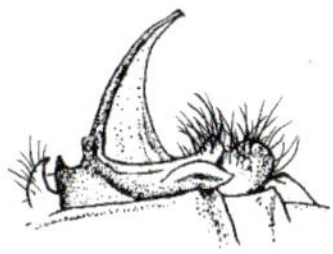

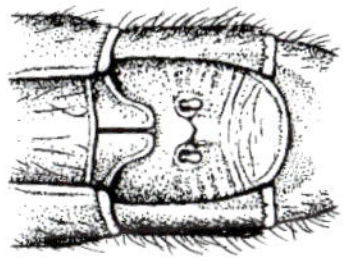

Belted Whiteface (pg. 246) *Leucorrhinia proxima*

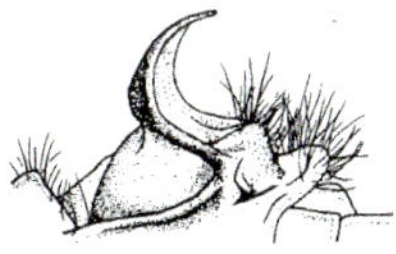

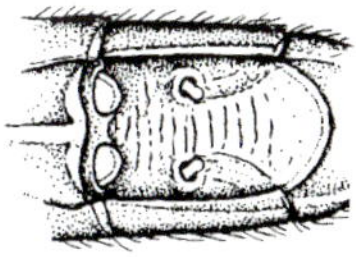

Crimson-ringed Whiteface (pg. 248) *Leucorrhinia glacialis*

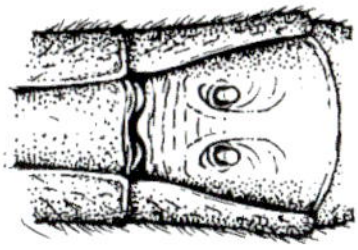

Boreal Whiteface (pg. 250) *Leucorrhinia borealis*

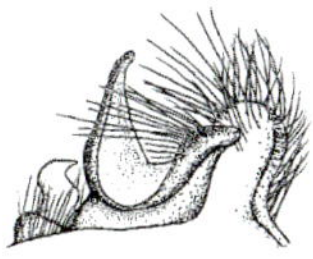

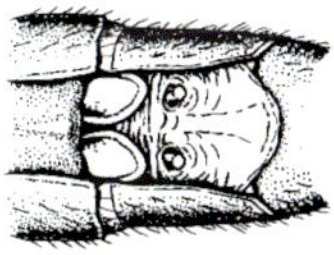

The Pondhawks — Genus *Erythemis*

This is a small genus of mid-sized, fiercely predatory skimmers known for taking large prey, including other dragonflies nearly their own size. Members have a series of three long spurs on the mid and hind femora (thighs) that are thought to be adapted for facilitating capture of large prey. Only one of the seven North American Pondhawk species occurs in the Rockies, the Western Pondhawk, but it has a closely related cousin that lives nearby. Eastern and Western Pondhawks are very similar and have at times been considered varieties of the same species, but are now regarded as distinct by most authors. Their ranges overlap slightly in eastern Colorado and New Mexico where some intergrading is probable. However, Rocky Mountain populations seem to be entirely the western species.

The Blue Dasher — Genus *Pachydiplax*

This genus consists of just one smallish North American species, the Blue Dasher, which is widely distributed across much of the country. It is probably most closely related to a genus of tropical dashers *Micrathyria*, several species of which are found in southern Texas. The characteristic that sets the Blue Dasher apart from all other odonates is a long, open cell with no crossveins below the stigma of each wing.

The Meadowhawks — Genus *Sympetrum*

The meadowhawks are a highly interesting genus of small, often abundant, late-season dragonflies famous for foraging on tiny insect prey in grassy meadows. Males are usually red or orange, often with black markings, and females are various shades of red, orange or brown. Tenerals are yellowish orange, gradually gaining adult coloration. No other Rocky Mountain genus is likely to be confused with this group, as other red, orange or brown dragonflies are larger. The Black Meadowhawk is an anomaly that could be confused with a whiteface because of its dark body, but it lacks the bright white face characteristic of the whitefaces. Behavioral characteristics of the group include a propensity to wander far away from water and a tendency to perch often and low on vegetation with wings drooped. Most species are not terribly wary, making them easy to observe. All of our species oviposit in tandem, at least part of the time; the only other reddish species to do so in the Rockies is the Red Saddlebags, which is much larger. This is an unusual dragonfly genus in that a number of species overwinter in a diapausing, middle-stage of egg development, with hatching occurring in the spring with the arrival of favorable temperatures and after spring rains fill their often temporary habitats. Meadowhawk eggs are large and the nymphs of most species complete development quickly, in just a few months in spring and summer. A number of species are frequently fliers at high elevations.

The genus has been subdivided into three groups, at various times referred to as categories, subgenera and even full genera. Of relevance to us, one of these groups has physical differences from the other groups that are useful in identification. I refer to this group as the *Tarnetrum* subgenus, which includes the first three species in the species accounts. All members of this subgenus (Variegated, Cardinal and Red-veined Meadowhawks) have a double row of cells in the radial planate region of the fore wing (illustration); the rest of our *Sympetrum* species usually have only a single row of cells there (rarely a few doubled cells in one wing).

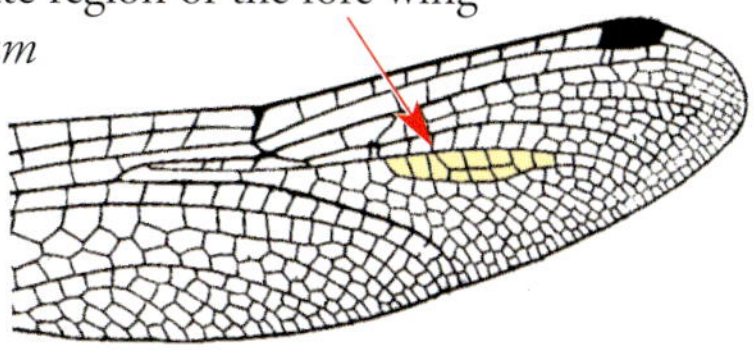

A cross-vein in the forewing's radial planate forms a double row of at least three cells in some species of *Sympetrum* (highlighted). Other species only have a single row of cells.

Meadowhawks may be the most difficult group of odonates in North America to identify to species. There are several factors that contribute to a good deal of confusion with this group. First, certain groups or pairs of species simply look very much alike. Second, most species give you a number of different "looks" because males and females are usually not similarly colored, and individuals of both sexes change appearance as they age. So, two species may look nearly identical at one stage of development, but completely different at another. Third, comparison of structural characters is often the only definitive way to separate some ages or sexes of some species, but differences may be slight, hard to see with a hand lens, and the there may be variation in the parts themselves. Last, hybrids and intergrades are far from unheard of with this group, although this is a bigger problem in the eastern U.S. than in the Rockies.

In short, this challenging group is a bit of a mess for the beginner to work through. Some of you will be content to simply call what you see a meadowhawk, and leave it at that, avoiding headaches and ulcers. Others, driven to know what everything is, will pursue the group in greater depth, and will eventually gain proficiency. For you, I leave the chart on pg. 256 (it describes the usual conditions, but exceptions occur) and the following suggestions to guide your approach for identifying our ten species. First, focus on identifying mature males, which are usually the easiest. Females and tenerals are not impossible to identify, but they can be tough, and are not the best places to start. Second, eliminate the easy species to reduce the possibilities. For example, Black Meadowhawks are the only non-reddish species. The mature male White-faced Meadowhawk is the only meadowhawk with a bright white face. Band-winged Meadowhawks are the only ones with clearly delineated basal orange patches on the hind wings. Autumn Meadowhawks are the species least marked with black and have pale-colored legs lacking black; further, the female is the only meadowhawk with a prominent, spout-like subgenital plate, easily visible in side view. Third, place the species you are observing into a group of similar species based on presence of white stripes on the sides of the thorax, black triangles on the sides of the abdomen, or pale veins on the leading edges of the wings. This should narrow the possibilities down to a small handful. Fourth, use range distribution maps to further narrow the possibilities. For example, if you're in the Central Rockies, you're probably not looking at a Cardinal Meadowhawk. Using known distributions of species in this way does not achieve iron-clad certainty—species can and do wander outside of their known ranges, and previously undetected populations do occur in unexpected places. However, currently understood ranges are worth cautiously weighing into the determination process. Fifth (and most important), make liberal use of your aerial net and your hand lens. Carefully note the coloration of the face, thorax, abdomen, legs and wing veins. Check the radial planate region of the fore wing for a double row of cells. Use the field characters given on the species pages in this guide, in conjunction with the chart on page 256 and the shapes of the hamular processes illustrated on pg. 257. This process should allow identification of even the toughest species. Enjoy the challenge! If it was easy, anybody could do it!

The Rainpool Gliders — Genus *Pantala*

This is a genus of large, strong-flying dragonflies with broad bases of the hind wings to permit near effortless gliding with the winds for long distances. They are brown, gray and yellow in color with a slender, tear-drop shape due to their large eyes which meet at a long eye seam on top of the head. They hang near vertically when perched, unusual for skimmers. Their superbly powerful and agile flight, and their far-ranging wanderings, often allows them to find the treasures of their existence—shallow, newly exposed or newly created pools of water for breeding—before other species. These select habitats typically lack fish and have few large insect predators. In these food-rich, low-risk environments, the nymphs have explosive growth rates, often completing development in four or five weeks. The nymphs are large, active and beautifully patterned, quite similar to those of saddlebags (Genus *Tramea*). There are just two species in this genus worldwide and both are found uncommonly in the Rockies, for the most part eschewing higher elevations, and occurring mostly on the western slopes of our regions. Both species are strongly migratory—capable of covering thousands of miles with few stopovers—and they share a very similar life history.

The Saddlebags — Genus *Tramea*

This strong-flying genus is similar to the rainpool gliders (Genus *Pantala*) in body shape, wing shape, large size and silhouette when in flight. The exceptionally broad hind wings are nearly as large as the rainpool gliders, at their base extending along about half the length of the abdomen. The prominent feature of many in the group, including both of our species, is the large, irregularly margined black or red patch near the base of the hind wing, presumably reminiscent of the saddlebags on a horse's saddle (hence the common name). These hind wing patches, along with the lack of any pruinosity on the body or wings, easily separates this genus from all other large dragonflies in the Rockies. They are powerful and agile fliers, feeding in sustained, gliding flight. Most, if not all species are apparently regular migrants, but embarrassingly little is known about the particulars of their movements. They droop their abdomens on hot days (opposite of obelisking), evidently to use shade afforded by their large hind wing patches. Mating pairs retreat to sheltered perches on vegetation near water and are infrequently seen. During tandem flight after mating, the female holds on to the abdomen of her male partner with her mid and hind legs. Males engage in interrupted guarding during oviposition (called "*Tramea*-type guarding," or TTG by Phillip Corbet) in which he releases the female when she descends from about a foot above water to the water surface to lay eggs, but grabs her again quickly when she ascends. These disengagements and reengagements are quick and smooth, done with much precision. The female sports a long, divided subgenital plate which she puts to good use holding clusters of eggs. The nymphs are large, attractively patterned and capable of very rapid growth; similar in those aspects to nymphs of rainpool gliders. Here in the Rockies we have records of two of the seven species known in North America, both infrequently reported and evidently rare here. However, both species are excellent dispersers and are capable of covering hundreds of miles in two shakes of a lamb's tail. So, keep your eyes open for them! Red Saddlebags and Black Saddlebags are similar in habits and hind wing pattern, differing primarily in color.

Desert Whitetail *Plathemis subornata*

APRIL	MAY	JUNE	JULY	AUG	SEPT	OCT

Breeding habitat: Hot springs and seepage pools in mostly arid landscapes. Habitat more varied south of Rockies.

Nature Notes:

Perches on tips of plants more often than on the ground.

Not nearly as well-studied as its close cousin, the Common Whitetail, with which it is rarely found in company.

Territorial defense and aggressive displays of males are thought to be similar to that species.

You can help increase knowledge about the behavior of Desert Whitetails and other poorly studied species by keeping careful notes of your observations!

Description: Adults average 1.7 inches long.

Male's thorax brown with muted pale spots on sides. Abdomen wide, compressed and covered completely with whitish pruinosity. Wings with dark streaks near bases, and wide, dark cross bands from near the nodus to the stigma when mature. White pruinosity covers most of wing from base to nodus. Head and eyes brown. Female's thorax with two yellow stripes on each side. Her wings also dark at bases but with two narrower, subapical, dark cross bands as though each of male's large bands had clear area in middle. Her abdomen brown with yellow lateral streaks on segments 1 to 8, forming interrupted stripes on each side. Immature male like female, including wing pattern; not so in Common Whitetail.

Identification Clues: Wings and abdomen distinctive. Male is most like Common

Dark wing patches of females are similar to those of teneral males (facing page).

Whitetail, but extensive white pruinosity on Desert's wings is reduced to small area on Common. Male Eight-spotted Skimmer has narrower black cross band and white pruinosity below stigma. Female and immature male easily distinguished from Common and Twelve-spotted Skimmer female by wings lacking black at tips, present on those species. Female Eight-spotted is larger, has white pruinosity on wings and only one black cross band near nodus.

Distribution: Scattered in southern $^{2}/_{3}$rds of Southern Rockies and southern tip of Central Rockies in Foothills and possibly lower Montane Zone.

Life Cycle & Behavior: Has long flight period with adults emerging in May and June and flying into autumn. Males perch along edges of breeding sites and actively defend territories. Behavior is less well known than Common Whitetail's, but likely similar. Nymph stage has been little studied.

Common Whitetail *Plathemis lydia*

APRIL	MAY	JUNE	JULY	AUG	SEPT	OCT

Breeding habitat: Quiet lakeshores; ponds, including cattle ponds; persistent puddles - all with mud bottoms.

Nature Notes:

Males adjust the intensity of their female-guarding behavior depending on the likelihood of harassment by other males.

In 1994 this species got me hooked on odonates. I found a road-killed male while bike riding and was struck by its beauty. I stuck it in my lunch box and brought it home to figure out what it was. My wife says I've haven't been the same since. I think she means that in a good way.

Description: Adults average 1.7 inches long.

Male's thorax brown with muted pale stripes on sides. Abdomen wide, compressed and covered completely with whitish pruinosity often having hint of light blue. Wings with black streaks near bases and wide black cross bands from the area of the nodus to the stigma. Small, triangular, white-pruinose patch on hind wing near base. Head and eyes brownish. Female's thorax brown with two yellowish stripes on each side. Her wings clear with dark blotches at bases, near nodus and at tips. Her abdomen brown with yellow lateral spots forming zigzag or zipper-like pattern on each side. Immature male with abdomen like female, but wing pattern like male; not so in Desert Whitetail.

Identification Clues: Wings and abdomen distinctive. Male is most like Desert Whitetail,

Note zig-zag or zipper-like pattern of yellow spots on female abdomen.

but lacks extensive white pruinosity present on Desert's wings. Male Eight-spotted Skimmer differs in having narrower black cross bands on wings and white pruinosity below stigmas. Females have black wing tips, lacking on Desert and Eight-spotted. Distinguish female from larger female Twelve-spotted Skimmer by lateral yellow abdomen spots, in-line on Twelve-spotted, zigzag or zipper-like pattern on Common.

Distribution: Foothills and Montane Zone of Northern Rockies and parts of Southern Rockies.

Life Cycle & Behavior: Has long flight period with adults emerging in late May and June and flying into autumn. Flies low and rapidly over water and perches frequently on the ground or other flat surfaces. Male aggressively defends territory using raised abdomen as threat display; submissive males lower theirs. Copulation is brief (3 seconds) and in flight. Female oviposits while hovering, tapping abdomen tip near clumps of vegetation or mud, releasing 25-50 eggs per tap, male usually hover-guarding nearby.

Chalk-fronted Corporal *Ladona julia*

APRIL	MAY	JUNE	JULY	AUG	SEPT	OCT

Breeding habitat: Lakes, ponds and slow streams, usually acidic, in open or forested landscapes.

Nature Notes:

Not afraid of people; many times this species has landed on my hat, shoulder, net or canoe paddle.

A highly social dragonfly—it is common to find dense congregations basking on horizontal surfaces in sunny places.

Description: Adults average 1.7 inches long.

Male's thorax is dark brown with a pair of thick whitish pruinose bars covering the top. Abdomen is short and black, with whitish pruinosity from base to segment 4. Eyes and face dark when mature. Wings mostly clear, with small dark markings at their bases. Females become colored like males when mature, but pruinosity usually darker, showing less contrast with body color. Immatures (tenerals) of both sexes orange brown with a thin black line along top of abdomen, lacking pruinosity, but corporal's bars still apparent (↑).

Identification Clues: Pair of wide, pruinose white bars on top of thorax, combined with mostly clear wings, easily separates this from all other Rocky Mountain species. Immatures lack pruinosity, but wide, light brown bars still easily visible. Partly pruinose abdomen of

Tawny tenerals (right) look much different when mature.

Chalk-fronted Corporals are highly gregarious and will bask in large groups.

matures also a field clue, but also present on some smaller whitefaces.

Distribution: Northern half of the Northern Rockies in Foothills and Montane Zone.

Life Cycle & Behavior: Adults emerge early, in late May and June and only occasionally fly into August. Frequently perches in groups, usually on the ground, logs or rocks, sometimes on low shrubs. From these perches its predatory forays are launched. Males fly low and swiftly over water, stopping occasionally to hover briefly. Females oviposit near shore by hovering, tapping, moving a short distance and repeating. Nymphs progress through about twelve molts and overwinter in final instar.

Four-spotted Skimmer *Libellula quadrimaculata*

APRIL	MAY	JUNE	JULY	AUG	SEPT	OCT

Breeding habitat: Lakes, ponds and some slow streams with soft bottoms, including bog ponds, often acidic.

Nature Notes:

No stranger to high elevations–breeds at 10,400 feet in a temporary pond in Colorado.

Has often been observed in migrating in huge swarms in Europe, not so here.

A Holarctic species widely distributed in Europe and Asia as well as this continent.

Scientific name means "four spotted," referring to the four nodal wing spots.

Description: Adults average 1.7 inches long.

Male's thorax brown with some yellow and black markings. Abdomen is compressed, light brown near base and narrowing and darkening toward the tip, with yellow streaks along sides of some segments. Wings usually marked with amber streaks along leading edges of all four (faint on some individuals) and small dark spots at each nodus. Hind wings each with a short amber streak at the base with a black triangle below it (↑). Head and eyes brownish. Female is same color and nearly same shape as male. Colors on both sexes become duller with age.

Identification Clues: Hind wing pattern is unique with amber at base with black triangle below it. Hoary and Bleached Skimmers superficially similar but with different hind wing pattern and not brown bodied. Immature

The amber-colored leading edges of all four wings are pronounced on this young female.

Often perches at the tip-top of branches.

Chalk-fronted Corporal is tawny brown but wing pattern dissimilar.

Distribution: All Rocky Mountain regions in Foothills, Montane and Subalpine zones.

Life Cycle & Behavior: Adults emerge very early, in May and June and fly into August. Immatures stray far from water after emergence. Perches on low vegetation, sometimes on ground. Voracious feeders on insects up to size of meadowhawks (genus *Sympetrum*). Males are highly territorial and aggressive, defending territories and (usually) egg-laying females from nearby. Females oviposit by hovering, tapping the abdomen tip repeatedly, moving a short distance and repeating the process. Nymphs progress through about 12 molts and overwinter in final instar.

Flame Skimmer *Libellula saturata*

APRIL	MAY	JUNE	JULY	AUG	SEPT	OCT

Breeding habitat: Wide variety of lakes, ponds, springs and slow streams; hot springs in north and at higher elevations.

Nature Notes:

I'll never forget how frustrating it was to try to net this species on a 103 degree F day in western Colorado. As the temperature increased they flew faster, but my swing got slower!

Dueling males may fly long, ascending, parallel flights coined "Red Baron" flights by odonatist Sid Dunkle.

Description: Adults average 2.1 inches long.

A conspicuously large, bright orange and well-named southwestern skimmer. Male's thorax and abdomen orange. Wings from base to nodus "saturated" with brownish orange color (hence scientific name) and with bright orange veins. Wings have short, brown basal streaks and their entire leading wing edges are amber even on females and immatures. Head and eyes orange on males, brownish on females. Female mostly brown with white line between wings and onto front of thorax. Her wings patterned similar to males but less colorful.

Identification Clues: No other Rocky Mountain species is very similar to either sex. Much larger than any orange meadowhawk.

Distribution: Parts of all the Rocky Mountain regions, mostly on western side of

Female is much paler than male on body and wings.

Continental Divide in Foothills and Montane Zone.

Life Cycle & Behavior: Adults emerge in June and fly into September. Males defend territories and aggressively chase other males (see Nature Notes). Mating is very brief (ten seconds) and while in flight. Male hover-guards female half-heartedly and only part-time. When female oviposits she sharply "flicks" her abdomen tip, spraying water and eggs forward into water or onto shoreline muck. Nymphs are hairy and sprawl on soft mud or silt bottom substrates.

Bleached Skimmer *Libellula composita*

APRIL	MAY	JUNE	JULY	AUG	SEPT	OCT

Breeding habitat: Alkaline marshes, brackish outflows from sink holes and spring seeps, especially hot springs.

Nature Notes:

One of the few king skimmers (along with Hoary) that lays eggs in tandem.

Maintains scattered, highly localized populations in alkaline or saline areas unsuitable for most odonates; often associated with rare Pupfish populations in desert Southwest.

The fragile habitats of this species are threatened by ground water depletion from well pumping in parts of its range.

Description: Adults average 1.8 inches long.

Male's thorax brown on top and pale on sides with a diagonal black stripe. Abdomen is striped with pale yellow and black; both thorax and abdomen turn dark as they become covered with grayish-blue pruinosity with age. Wings are distinctive, each with a pale costal vein along the leading edge, a modest amber-brown patch at the base, a small brown dot at the nodus and a black stigma. Face whitish; eyes light brown. Female's thorax is pale with dark stripes on top, at shoulder and on side. Her abdomen contrastingly pale and black striped. Her wings like male's, but usually lacking dark dots at nodus.

Identification Clues: Easily distinguished in field when perched. Wing pattern described above is unique and pale face is hard to miss.

Wings of this young female have faint dark patches at the bases.

Other somewhat similar king skimmers lack pale costal vein on leading wing edges and have larger dark blotches on wings.

Distribution: Rare in southern tip of Central Rockies of Utah in Foothills and perhaps lower Montane Zone.

Life Cycle & Behavior: Adults emerge in June and fly into August. Males energetically patrol water edges, perching occasionally on tips of reeds, low shrubs or on ground. Female wraps legs around male's abdomen in tandem flight. Pair oviposits in tandem, rising and dropping when female repeatedly taps and splashes water surface with abdomen to disperse eggs. Little is known about the nymph stage.

Hoary Skimmer *Libellula nodisticta*

APRIL	MAY	JUNE	JULY	AUG	SEPT	OCT

Breeding habitat: Ponds, slow streams and springs, especially hot springs in arid landscapes.

Nature Notes:

Females sometimes oviposit in tandem but usually oviposit alone.

When alone, a female deposits her eggs with several quick swings downward from 2 feet above water, tapping her abdomen tip just once for each swing.

Common name alludes to grayish pruinosity of mature adults thought to resemble hoarfrost.

Description: Adults average 2.0 inches long.

Male's thorax dark brown with pairs of yellow spots on each side. Abdomen dark brown with yellow, interrupted dorsolateral stripes on each side, diverging away from thorax. Both thorax and abdomen becoming bluish gray pruinose with maturity. Wings distinctive, each with dark spots of moderate size at base and nodus, basal spots elongate with light pruinosity surrounding. Face and eyes brown; forehead black. Female colored and patterned like immature male; may also develop pruinosity on body when old. Her face brownish yellow and forehead yellow.

Identification Clues: Easily identified in the field when perched by the pattern of dark spots on wings described above, which are unique for both sexes. Lacks the pale costal

Note moderate-sized basal and nodal wing patches on this female.

wing veins of the slightly smaller Bleached Skimmer. Dark wing spots are smaller than those of other king skimmers with which it is most likely to be confused.

Distribution: Rare in southern half of Southern Rockies and southern tip of central Rockies in Foothills and lower Montane Zone.

Life Cycle & Behavior: Adults emerge in June and fly into August. Males intermittently patrol edges of breeding sites, perching often on tips of plant stems in open, launching feeding forays from there. Female usually oviposits alone (see Nature Notes), but pairs occasionally oviposit in tandem, joining Bleached Skimmer as only king skimmers to do so. Female disperses eggs in typical king skimmer fashion by tapping water surface near vegetation with her abdomen. Little is known about the nymph stage.

Eight-spotted Skimmer *Libellula forensis*

APRIL	MAY	JUNE	JULY	AUG	SEPT	OCT

Breeding habitat: Wide variety of vegetated ponds, slow streams and lakeshores; often larger lakes.

Nature Notes:

Males often perch on reed tips and other exposed locations near shore, thus easy to observe.

Common name refers to the total number of dark wing spots.

Known to have hybridized with the Widow Skimmer.

Description: Adults average 2.0 inches long.

A large, remarkably patterned, western species. Male's thorax dark brown with white pruinosity on top and a pair of yellow spots low on each side. Abdomen top completely white pruinose when mature. Wings distinctive, each with a prominent dark streak at base and a dark crossband just past the nodus. Patches of white pruinosity on wings margin basal dark streaks and extend below stigma. Face and eyes of both sexes brown. Female's thorax like male's but lacking top pruinosity. Her wings like males, but with less white pruinosity, sometimes none. Her abdomen dark brown with yellow, interrupted dorsolateral stripes on each side, often developing pruinosity when old.

Identification Clues: Wings and abdomen distinctive. Male has narrower black cross

No other female king skimmer has wings patterned anything like the female Eight-spotted.

bands on wings than either species of whitetail. Male most like Twelve-spotted Skimmer but lacks black at wing tips, almost always present on Twelve-spotted. Females lack black wing tips, present on Twelve-spotted and Common Whitetail females. Females usually have white pruinosity on wings, lacking on other female skimmers.

Distribution: All Rocky Mountain regions in Foothills and Montane Zone.

Life Cycle & Behavior: Adults emerge in June and may fly into September. Males engage in lengthy patrols over large areas, not defending fixed territories, but behave aggressively toward any dragonfly males of similar size. Female disperses eggs by tapping water surface with her abdomen over large or small area, sometimes with male hover-guarding nearby. Nymphs over-winter in a late instar and may crawl some distance from water to transform.

Twelve-spotted Skimmer *Libellula pulchella*

APRIL	MAY	JUNE	JULY	AUG	SEPT	OCT

Breeding habitat: Moderately vegetated lakes, ponds and slow streams; especially in open areas exposed to sun.

Nature Notes:

One of the highest-elevation odonates, found at 10,700 feet in Colorado.

Often returns to the same bare, prominent twig-tip perch when flushed from it.

Common name refers to the total number of dark wing spots.

Scientific name *pulchella* means "pretty," no doubt referring to the coloration of the male.

Description: Adults average 2.1 inches long.

A large, handsome, transcontinental species. Male's thorax brown with 2 yellowish stripes on each side. Abdomen brown with yellow lateral stripe on each side, becoming grayish pruinose with age, but not completely so. Wings distinctive, each with three prominent dark spots: at base, near nodus and at tip. Patches of white pruinosity on hind wings at bases along abdomen and between dark patches on all wings. Face and eyes of both sexes brown. Female similar to male, but her abdomen not turning pruinose, showing continuous, yellow lateral stripes. Her wings with dark spots like male but lacking white pruinosity. Rare individuals lack dark spots at wing tips.

Identification Clues: Unmistakable. No other king skimmer has three prominent dark spots on each wing. Eight-spotted Skimmer

Note in-line pattern of lateral spots on the abdomen of this female. Contrast with zipper-like pattern on abdomen of Common Whitetail (inset).

Prefers tip-top perches.

lacks dark spots at wing tips. Female similar to female Common Whitetail, but is larger and has lateral yellow abdomen spots in-line and appearing continuous; female Common has abdominal spots in zig-zag or zipper-like pattern (↑).

Distribution: All Rocky Mountain regions in Foothills, Montane and Subalpine zones.

Life Cycle & Behavior: Adults emerge in late June and July and may fly into September. Males make long, fast swoops over ponds, stopping abruptly to hover, then dart off. Males are very aggressive toward other male dragonflies and defend a shifting territory. Female oviposits, often alone, near submerged vegetation by rhythmically yet vigorously striking the water surface with her abdomen, spraying eggs and water droplets up to a foot in front of her. Nymphs are large and hairy; overwinter in a late instar.

Widow Skimmer *Libellula luctuosa*

APRIL	MAY	JUNE	JULY	AUG	SEPT	OCT

Breeding habitat: Wide variety of lakes, ponds and slow streams with soft bottoms in open or forested landscapes.

Nature Notes:

Often launches feeding forays from plant tip perches in open fields.

Known to have hybridized with the Eight-spotted Skimmer.

Scientific name *luctuosa* means "mournful," perhaps alluding to the dark, cloak-like basal wing patches. Common name evidently based on same allusion.

Description: Adults average 1.8 inches long.

This skimmer must have invented the term "showy." Male's thorax brown with white pruinosity on top. Abdomen slender and thinly covered with whitish gray pruinosity. Wings diagnostic, each with huge basal dark patch out to nodus and white pruinose from there nearly to stigma. Head and eyes dark brown. Female has similar and diagnostic wing pattern, though dark basal patches less extensive and lacking white pruinosity; small blackish tips to wings. Her abdomen black on top with wide yellowish brown lateral stripes. Immature males resemble females.

Identification Clues: Unmistakable. No other Rocky Mountain species has wing pattern remotely like it. Black Saddlebags has large basal wing patch on hind wing only, its margins irregular and with no white pruinosity

Huge basal wing patches on this species are impossible to miss!

on wings. Other king skimmers with black at wing bases have much smaller patches there.

Only the males of this species develop the striking wing pruinosity.

Distribution: East Slope of the Southern Rockies and southern tip of Central Rockies in Foothills and Montane Zone.

Life Cycle & Behavior: Adults emerge primarily in June and fly into September. Males are aggressive toward other male dragonflies and generally fly short beats to defend individual shifting territories. At high densities, group territories may be established and only dominant males mate. Females disperse eggs by tapping water surface with abdomen in open water with males sometimes hover-guarding nearby. When other males are few, males may not bother to guard. Nymphs overwinter in a late instar.

Dot-tailed Whiteface *Leucorrhinia intacta*

APRIL	MAY	JUNE	JULY	AUG	SEPT	OCT

Breeding habitat: Small ponds, marshy bays and slow streams, usually shallow with emergent vegetation in open areas.

Nature Notes:

One of the first dragonflies to emerge in spring throughout its range.

The ability of this species to disperse was found to be negatively correlated with water mite parasitism.

Known to hybridize with Crimson-ringed Whiteface.

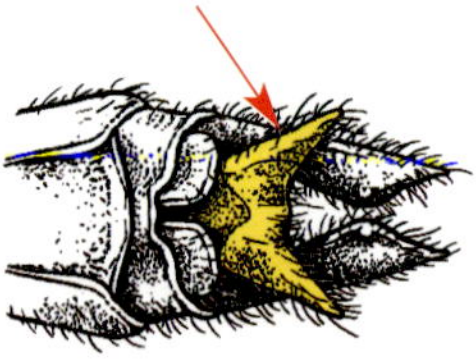

Ventral view of male's epiproct showing widely divergent tips. Other whitefaces have more parallel-sided epiprocts.

Description: Adults average 1.3 inches long.

Male's thorax and abdomen black, marked only with prominent, squarish yellow spot on top of segment 7 (↑). Epiproct widely divergent toward tip (↑). Eyes dark brown, face ivory white. Wings clear, but some females with various amounts of amber wash at base. Female's thorax brown. Her abdomen brown with yellow along sides of segments 3 to 5, and yellow streaks on top of middle abdominal segments, ending with a short, squarish spot on top of segment 7. Her subgenital plate with pair of slender projections like stubby thumbs (pg. 217). Tenerals of both sexes with much yellow on abdomen top and between wings on thorax, most darkening and disappearing with age.

Identification Clues: Mature male is unmistakable with all black body and squarish yellow spot on segment 7. Females have various

The pattern of spots on this teneral male are similar to those on females.

amounts of yellow on top of abdomen depending on maturity, but spot on segment 7 always squarish and shorter than on most other whitefaces. Teneral males and females have diagnostic reproductive parts seen with hand lens as described above (pg. 217).

Often perches on the ground.

Distribution: All Rocky Mountain regions in Foothills and Montane.

Life Cycle & Behavior: Adults emerge early—May and early June in most areas—and fly into August. Often perches on ground in sunny openings. Obelisks to cool on hot days. Males intermittently defend modest territories (up to 12 feet in diameter) from perches on water lilies or emergent stems often away from shore. Female oviposits by tapping water surface irregularly over submerged vegetation while in swift flight, usually with male hover-guarding above. Nymphs not currently separable from Belted Whiteface.

Hudsonian Whiteface *Leucorrhinia hudsonica*

APRIL	MAY	JUNE	JULY	AUG	SEPT	OCT

Breeding habitat: Pools in open bogs and fens; shallow, cold, marshy ponds.

Nature Notes:

A true high-elevation warrior—In 2007, Scott Wissinger showed me a population at 9,500 feet near the southern tip of its range in Colorado.

Early flight period in boggy areas often coincides with emerging hoards of mosquitoes, on which they feed heavily.

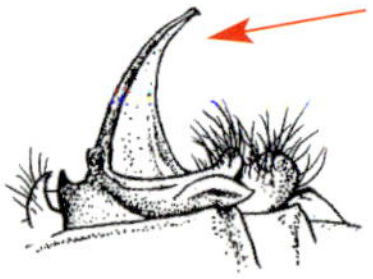

Male hamular process (in this view, upside down; located under abdomen near thorax).

Description: Adults average 1.2 inches long.

A small whiteface with a conspicuously spotted abdomen. Male's thorax red with black markings on top and sides. Abdomen black with bright red at base (segments 1 to 3) and red spots on segments 4 to 7. Eyes dark brown, face ivory white. Typical female (heteromorph) yellowish where male is red with abdomen spots more elongated; male-like female (andromorph) just like male. Her subgenital plate about half the length of segment 9, with elongate, triangular lobes parallel and adjacent along inner margins. Red areas are yellow on tenerals.

Identification Clues: Small size and conspicuously spotted abdomen separate it from most other whitefaces. Boreal Whiteface most similar but considerably larger and has red spot on segment 8, lacking on Hudsonian.

A typical yellow-form female.

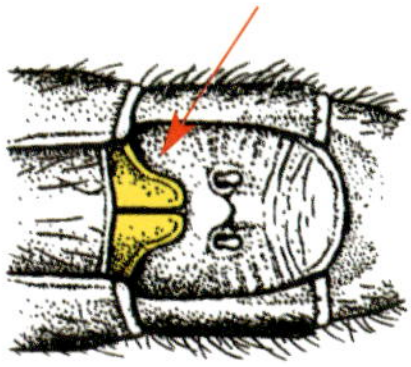

Lobes of female subgenital plate parallel-sided and adjacent (highlighted).

Hamular processes also distinctive (↑) (pg. 217). Female best identified in the hand by the elongate lobes of subgenital plate parallel and adjacent along the inner margin of their length (↑) (pg. 217); lobes diverge toward tips on Boreal and variously different on other whitefaces.

Distribution: All Rocky Mountain regions in Montane and Subalpine zones.

Life Cycle & Behavior: Adults have early but long flight period in Rockies, emerging in May and June and flying into August. Perches on low vegetation, logs or ground. Males intermittently defend small territories of about one square yard. Females oviposit by quickly tapping abdomen when hovering, often near sedges or sphagnum moss with male hover-guarding overhead. Nymphs are small with three easily seen longitudinal black lines on the underside of the abdomen.

Belted Whiteface *Leucorrhinia proxima*

APRIL	MAY	JUNE	JULY	AUG	SEPT	OCT

Breeding habitat: Heavily vegetated ponds and lakeshores, often boggy, in forested areas.

Nature Notes:

A high-elevation whiteface recorded at over 9,300 feet in Colorado.

Males in eastern U.S. with yellow rather than red markings, and East of Rockies males develop pruinosity at base of abdomen (segments 1-4), not so here.

Formerly known as the Red-waisted Whiteface.

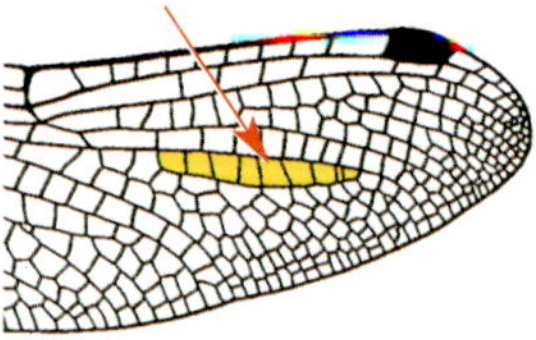

Radial planate of forewing has a single row of cells (highlighted). This trait separates the Belted from the Crimson-ringed, which has a double row of at least three cells here.

Description: Adults average 1.4 inches long.

Male's thorax black with areas of bright red on top. Abdomen black with bright red at base (segments 1 to 3) and often with thin red streaks on top of segments 4 to 7. Hamules with inner branch smoothly curved and outer branch lacking (pg. 217). Eyes brown, face ivory white. Typical female (heteromorph) yellow where male is red; male-like female (andromorph) with red markings. Her subgenital plate short with rounded lobes more than twice as wide as long. Red areas are yellow on tenerals.

Identification Clues: Both sexes are most similar in size and color to Crimson-ringed Whiteface, but male hamules lacking outer

Degrees of pruinosity in males varies from none to nearly obscuring the dark abdomen.

branch are distinctive (compare on pg. 217). Both sexes have two or fewer double cells (usually single row only) in radial planate region of the forewing (↑); Crimson-ringed has at least three doubled cells there. Female subgenital plate with lobes very short but evident, even shorter on Crimson-ringed; other whitefaces with noticeably longer lobes (pg. 217).

Distribution: Parts of all Rocky Mountain regions in Foothills, Montane and lower Subalpine zones.

Life Cycle & Behavior: Adults emerge in late May and June and fly into August. Males defend small territories from perches near shore (one to two square yards). Mating is fairly lengthy (> 15 minutes), occurring while perched on shrubs or trees. Male may employ sly karate-guarding strategy (see genus summary). Female taps water surface while in flight to release eggs. Nymphs not currently separable from Dot-tailed Whiteface.

Crimson-ringed Whiteface *Leucorrhinia glacialis*

APRIL	MAY	JUNE	JULY	AUG	SEPT	OCT

Breeding habitat: Lakes and ponds, usually boggy or marshy, in forested landscapes.

Nature Notes:

Called a "singularly beautiful insect" by pioneering odonatist James Needham in 1901.

May be locally abundant at prime habitats.

Nymphs have three conspicuous, longitudinal black lines on the underside of the abdomen, similar in that regard to Hudsonian and Boreal Whitefaces.

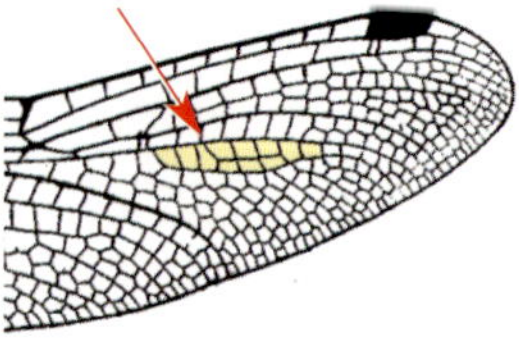

A cross-vein in the forewing's radial planate forms a double row of at least three cells (highlighted). This trait separates the Crimson-ringed from the Belted, which has a single row of cells here.

Description: Adults average 1.4 inches long.

Male's thorax black with bright red markings on top and sides. Abdomen black with red at base (segments 1 to 3) and sometimes with thin red streaks on top of segments 4 to 7. Hamules unique, looking like gaping alligator jaws from side. Eyes dark brown, face ivory white. Typical (heteromorph) female yellowish where male is red; male-like female (andromorph) with red markings. Her subgenital plate extremely short, appearing as pair of tiny bumps. Red areas become duller with age, yellowish on tenerals.

Identification Clues: Both sexes are most similar in size and color to Belted Whiteface, but male hamules described above are diagnostic with hand lens (pg. 217). Both sexes have at

This teneral male is yellow, but he'll turn red when he matures.

least three double cells in radial planate region of the fore wing (see illustration) (↑); Belted has two or less double cells there, usually only single row. Female has shortest lobes on her subgenital plate of any whiteface, barely noticeable (pg. 217).

Distribution: Uncommon in Foothills and Montane Zone in northern areas of Northern and Central Rockies.

Life Cycle & Behavior: Adults usually emerge in June and fly into early August. Often perches flat on ground in sunlight or shade, but sometimes on shrubs or floating vegetation, from there launching feeding attacks. Males defend small territories.

Females oviposit by quickly tapping abdomen when hovering, often with male hover-guarding overhead.

Boreal Whiteface *Leucorrhinia borealis*

APRIL	MAY	JUNE	JULY	AUG	SEPT	OCT

Breeding habitat: Sedge marshes, bog and prairie ponds, sloughs and lakeshores, usually in open landscapes.

Nature Notes:

Ecology and life history in U.S. Rockies is poorly known because of its spotty, localized distribution, so record your observations!

May be abundant at prime habitat.

Some northern species occur only in high elevation regions (Montane and Subalpine zones) when the southern limit of their ranges is in the Rockies. Such is the case with the Boreal Whiteface in the Uinta Mountains.

Very similar in appearance to its close relative the Hudsonian Whiteface, only larger.

Description: Adults average 1.5 inches long.

Our largest whiteface and the most heavily marked with red or yellow. Male's thorax with complex markings of red, brown and black. Abdomen black with large red markings on segments 1 to 8. Eyes dark brown, face ivory white. Typical (heteromorph) female yellowish where male is red; male-like female (andromorph) with red markings. Both female forms with large pale spot on segment 7, but usually no spot on segment 8. Her subgenital plate with two large lobes rounded at tips.

Identification Clues: Colored most similarly to Hudsonian Whiteface, but is much larger than Hudsonian and with more red on abdomen including segment 8 (↑). Female differs from other whitefaces of almost same size in having large spot on segment 7 (↑), lacking on Belted and Crimson-ringed Whitefaces.

This red-form female is colored much like the male.

Confirm identifications in hand by comparing male hamules and female subgenital plates with similar species (pg. 217).

Distribution: Local in Montane Zone of Central and Southern Rockies.

Life Cycle & Behavior: Flight period is early and short, with adults emerging in late May and June and flying only into July. Males perch most commonly on upright stems of emergent vegetation over water, but also on floating vegetation, trees and the ground. Territorial and reproductive behavior thought to be similar to Hudsonian Whiteface.

Western Pondhawk *Erythemis collocata*

APRIL	MAY	JUNE	JULY	AUG	SEPT	OCT

Breeding habitat: Ponds, lakeshores, slow streams and marshes; also pools at hot springs in Northern Rockies.

Nature Notes:

The Western Pondhawk is a fierce predator on damselflies, moths and other large insects, but is not known to cannibalize other pondhawks as its eastern cousin does.

During their reproductive period, females may lay eggs once a day and appear obsessed with sex, likely mating at least once a day!

Description: Adults average 1.6 inches long.

Mature male's body entirely light blue pruinose, often with some green showing through sides of thorax. Eyes greenish blue, face green. Female's abdomen and thorax green with fine black line down middle of abdomen on segments 2 to 9. Some populations have black more extensive on female's abdomen. Her eyes green. Terminal appendages dark on male (↑), pale on female. Tenerals of both sexes bright green.

Identification Clues: Eastern Pondhawk unlikely to occur in Rockies; male Western differs from Eastern in having dark cerci (white on Eastern) and a shorter, stouter abdomen. Male is superficially similar to smaller Blue Dasher, but Blue Dasher has white face, dark streaks at wing bases, a long, open cell below

Female's green abdomen with black line down the middle is distinctive.

each stigma and striping on thorax usually showing through pruinosity, all lacking or different on Western. Female's coloration unique among skimmers in Rockies.

Distribution: All Rocky Mountain regions in Foothills and Montane Zone.

Life Cycle & Behavior: Adults have a long flight period from May through September. Perches and flies low. Females oviposit by tapping infrequently in one spot, then moving to another, while males guard by flying around her, not hovering. Nymphs are quite distinctive, cute actually, with stocky builds and down-turned tips of the paraprocts.

Blue Dasher *Pachydiplax longipennis*

APRIL	MAY	JUNE	JULY	AUG	SEPT	OCT

Breeding habitat: Quiet weedy ponds; also vegetated lake bays, ditches, marshes and very slow streams.

Nature Notes:

Male performs aerial displays in front of female after mating to lead her to the egg-laying site.

Roosts in trees at night, but evidently attracted to light; many have been taken in light traps between 9pm and dawn.

Often perches on twigs and stem tips with wings drooped, possibly to shade thorax; also obelisks on hot days (see photo pg. 26).

Description: Adults average 1.4 inches long.

Mature male's body entirely light blue pruinose, often with some striping showing through on thorax and with dark abdomen tip sometime apparent. Eyes bright green, reddish brown on tenerals; face creamy white. Wings with dark streaks at bases; often other amber markings near base and beyond nodus, but age-related and not constant across range. Female's abdomen black with two distinctive, longitudinal yellow stripes except at tip (↑). Her thorax heavily striped brown and yellow, as is immature male's.

Identification Clues: Male similar to larger Western Pondhawk, but the Blue Dasher's white face (green on Western) and striping on thorax usually showing through pruinosity (lacking on Western) are good field characters. Male may have dark abdomen tip, good clue

Note unique double row of spots on top of female's abdomen (inset).

when present. Female's heavily striped thorax and double row of yellow spots on abdomen are good field marks. In the hand, the long, open cells below each stigma, 3 or 4 times longer than nearby cells (illustration), are diagnostic for both sexes.

Note long, open cell below stigma (highlighted), unique to Blue Dasher.

Distribution: Uncommon in scattered parts of all Rocky Mountain regions in foothills and Montane Zone.

Life Cycle & Behavior: Has long flight period, emerging in June and persisting well into autumn. Males defend near-shore territories, hovering often, darting and dashing here and there to snatch prey or confront intruders. Both sexes will defend favored feeding perches and are estimated to catch about 300 tiny prey insects a day. Dueling males confront head to head and raise their abdomens in threat displays, each trying to get lower than the other to force opponent up and away from water. Female oviposits by hovering near water over submerged vegetation, repeatedly dunking her abdomen tip, taking only about 30 seconds to release 300 to 700 eggs. Male guards by perching or hovering overhead. Nymphs mature in one year and overwinter in later instars.

Meadowhawk Identification Chart

When two or more states are described, the first state describes the mature male; the second describes females and tenerals.

	Face color	Upper leg (thigh) color	Thorax markings
Variegated	Red; tan	Black and pale striped	Yellow spots below; pale stripes leading to yellow spots below
Cardinal	Red; brown	Red or tan	Yellow spots below
Red-veined	Red; brown	Black	None; pale stripes
White-faced	White; greenish	Black	None
Striped	Tan	Variable	Pale stripes
Cherry-faced	Red; brown	Black	None
Saffron-winged	Red; brown	Striped black and pale	None
Band-winged	Tan	Black	Black striping
Black	Black	Black	Black; complex black and pale pattern
Autumn	Red; brown	Brown, orange or yellow	None

	Abdomen markings	Wing markings
Variegated	Multicolored with complex pattern	Costal margin red or yellowish
Cardinal	Unmarked red or brown	Costal margin and base reddish
Red-veined	None; black and white lateral stripes	Costal margin and base reddish
White-faced	Black lateral triangles	None
Striped	Black lateral triangles, may be reduced	None
Cherry-faced	Black lateral triangles	Light yellow wash often with some orange at base
Saffron-winged	None; thin black lateral lines	Costal margin red or yellowish
Band-winged	Black lateral triangles; reduced to black lines	Orange-brown basal patches
Black	None; black and white lateral stripes	None
Autumn	None	None

	Hamules (see illustrations on facing page)
Variegated	Drooped type
Cardinal	Drooped type
Red-veined	Mitton type
White-faced	Modified mitten
Striped	Modified mitten
Cherry-faced	Deeply forked for 1/3rd of length
Saffron-winged	V-shaped and stubby
Band-winged	V-shaped
Black	V-shaped
Autumn	V-shaped

Generalized male hamule (lateral view)

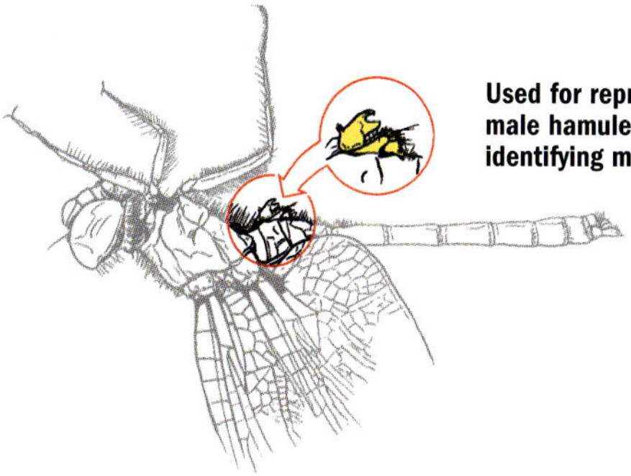

Used for reproduction, the male hamule can help in identifying meadowhawks.

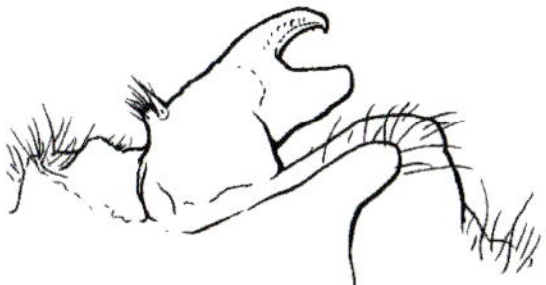

Forked-type hamule

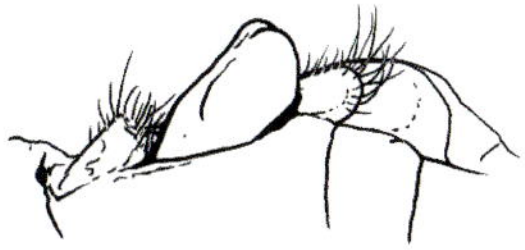

Mitten-type hamule

V-shaped hamule

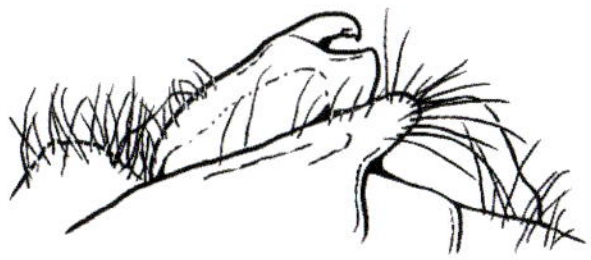

Modified mitten-type hamule

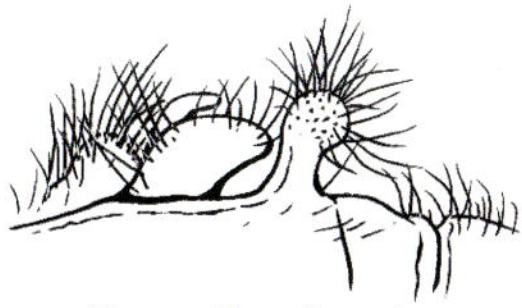

Drooped-type hamule

Variegated Meadowhawk *Sympetrum corruptum*

APRIL	MAY	JUNE	JULY	AUG	SEPT	OCT

Breeding habitat: Sand bottomed ponds, shallow, marshy lakes and slow streams, often alkaline, saline or stagnant.

Nature Notes:

Has been found at 13,000 ft. in Colorado, but unlikely to breed that high. High altitude species worldwide often fly higher than their breeding sites and extreme weather events can take individuals up into unfamiliar territory.

A wary species, hard to approach and challenging to net.

Migrates in spring from south to north and irregularly wanders from west to east; migrating swarms sometimes seen.

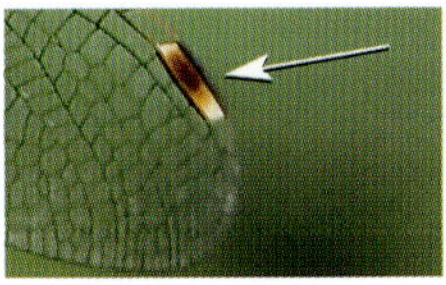

Description: Adults average 1.5 inches long.

Mature male's thorax brown with pairs of small yellow spots low on sides. Abdomen shows a complex color pattern, mostly red with areas of gray and sometimes brown and with a series of white lateral spots fading with age. Eyes, face and leading-edge wing veins reddish. Two-toned stigmas darker in middles than at edges (↑). Legs black with yellow stripes. Female's thorax brown with two persistent white side stripes leading down to yellow spots. Her abdomen also complex, brown (↑) where male is red, and white side spots on segments 3 to 8 persisting. Her eyes and face tan; leading-edge wing veins yellowish.

Identification Clues: Similar to several meadowhawks with red/orange/yellow veins on leading edges of wings, but shares yellow spots on sides of thorax only with the Cardinal

Note the complexly patterned abdomen of this female.

Meadowhawk, and the abdomen is more complexly patterned than any other species. Cardinal is brighter red, has dark markings at hind wing bases and lacks any pattern on abdomen. Other meadowhawks with white side stripes lack yellow spots at the lower ends. Also note striped legs, drooped-type hamule (pg. 257) and stigma color not uniform.

Distribution: All Rocky Mountain regions in Foothills, Montane and Subalpine zones.

Life Cycle & Behavior: Flies all year in southern U.S.; arrives in north in spring, prior to any sign of emergence. Nymphs emerge late in summer or fall in north and soon leave. Flies low over water, stopping often to perch on twigs, low vegetation or ground. Often found far from water. Ovipositing pairs stay low, tapping repeatedly in one spot, then move elsewhere and repeat. Nymph stage, as well as migratory strategy, could use more study. Tolerates stagnant, malodorous habitats humans have termed "nasty."

Cardinal Meadowhawk *Sympetrum illotum*

APRIL	MAY	JUNE	JULY	AUG	SEPT	OCT

Breeding habitat: Lakes and pond in northern range; elsewhere small ponds and pools in streams.

Nature Notes:

Females are reclusive, often hard to find, unusual for meadowhawks.

Canadian entomologist Francis Whitehouse once exclaimed, "what more striking contrast in nature than a number of bright scarlet *illotum* flying over green reeds!"

Description: Adults average 1.4 inches long.

Mature male's entire body brilliant red. Thorax with pairs of pale yellow spots on lower sides. Abdomen is broad and parallel-sided. Eyes, face and legs mostly red. Veins along leading edges and at bases of wings reddish; brown streaks at bases of wings, most noticeable on hind wings. Female much like male—either red or brown—and developing whitish pruinosity on underside of abdomen with age.

Identification Clues: Brilliant red males are hard to misidentify. The combination of solid red abdomen, paired yellow spots on lower sides of thorax, brown streaks at wing bases and reddish veins on leading edges of wings is diagnostic. Drooped-type hamule distinctive in side view (pg. 257). Red-veined Meadowhawk is similar but lacks brown basal wing streaks and conspicuous yellow side spots on thorax

This reddish female is colored much like a mature male.

(has faint stripes there). Same combination of characters distinguishes females from other female meadowhawks.

Distribution: Uncommon in scattered areas of Northern Rockies in Idaho and Southern Rockies in New Mexico.

Life Cycle & Behavior: Adults emerge in late May and June and fly into September. Perches on low vegetation near or over water with drooped wings, returning repeatedly to favored perches; obelisks on hot days to cool. Males defend shoreline territories and harass intruders of any species. Ovipositing pairs stay close to shore, tapping near emergent vegetation or algal mats while hovering; females may continue alone after male tires and takes off. Life history of the nymph is poorly known.

Red-veined Meadowhawk *Sympetrum madidum*

APRIL	MAY	JUNE	JULY	AUG	SEPT	OCT

Breeding habitat: Shallow, marshy, temporary ponds, sometimes saline, with dense emergent vegetation in open landscapes.

Nature Notes:

His even-textured red-veined wings make the male conspicuous to the observer.

Wary and challenging to approach.

Scientific name *madidum* means "wet" or "moist," but like many species named by Hermann Hagen, the allusion is unknown.

Description: Adults average 1.6 inches long.

Mature male's entire body brilliant red. Thorax with pairs of faint pale spots on lower sides. Abdomen may have tiny black marks on sides and near tip. Eyes and face red; legs black. Veins along leading edges and at bases of wings reddish. Females light brown, sometimes red, with two wide white stripes on sides of thorax. Her abdomen has thin lateral side stripes on segments 4 to 9 and white below them. Her subgenital plate about 1/3rd as long as segment 9 and V-notched.

Identification Clues: Cardinal Meadowhawk male is most similar but has brown streaks at wing bases—lacking on Red-veined—and hamules differ between species, those of Red-veined mitten-like (pg. 257). Striped thorax of female and teneral Red-veined much like Striped Meadowhawk, but the wings of Red-

Female is very similar to female Striped Meadowhawk, but has doubled row of cells in radial planate region of forewing (see illustration, pg. 219).

veined are more reddish at bases, the mitten-type hamule is dissimilar between the species and Red-veined has a doubled row of cells in radial planate region of fore wing, not so on Striped.

Distribution: Sparsely distributed in Foothills and Montane Zone of Northern and Central Rockies; rare and local in Southern Rockies.

Life Cycle & Behavior: Adults emerge primarily in June and fly into September, but there may be a secondary emergence peak in fall in some areas. Perches on tips of low plant stems or suns on ground. Females oviposit in tandem or alone, near emergent plants in shallow water, sometimes on dry pond bottoms. Male often disengages during egg-laying and hovers or perches nearby. Evidently overwinters in egg diapause. Little is known about the life history of the nymphs.

White-faced Meadowhawk *Sympetrum obtrusum*

APRIL	MAY	JUNE	JULY	AUG	SEPT	OCT

Breeding habitat: Ponds, often temporary; also shallow marshes, bogs and fens, often in forested landscapes.

Nature Notes:

I once watched numerous pairs oviposit on the grass of a soccer field on top of a small hill with no water nearby. Evidently they mistook the level surface and short-cropped, recently watered grass for the bottom of a temporary pond!

Often abundant in wet sedge meadows and other prime habitats.

Description: Adults average 1.3 inches long.

Male's thorax unmarked brown. Abdomen bright red with black, forward-pointing triangles on segments 4 to 9. Face pearly white; eyes reddish brown; legs black; and terminal appendages reddish. Wings clear and dark-veined but sometimes with touch of amber at bases (never extensive). Female has markings like male, but is light brown instead of red, brown darkening with age. Her eyes and face brownish. Her subgenital plate about half length of segment 9, with two tips pointing rearward. Tenerals like female but yellow-orange, with yellowish green on face and lower eyes.

Identification Clues: Mature male is unmistakable with bright white face, red abdomen and black legs. Teneral males similar to several other meadowhawks, but clear wings, unmarked thorax, extensive black on abdomen

Brownish female contrast markedly with the red male in this mating pair.

and shape of modified, mitten-type hamule in side view (pg. 257) allow identification. Female has darker wings veins than otherwise similar female Cherry-faced Meadowhawk, from which it cannot always be distinguished. Slight differences exist in shapes of subgenital plates between the two species, but with some overlap. Females become darker brown with age than other meadowhawks.

Distribution: All Rocky Mountain regions in Foothills, Montane and possibly lower Subalpine zones.

Life Cycle & Behavior: Adults have a long flight period, emerging in late June and July and flying into October. Female oviposits in shallow water, on mud, on bank-side grasses or even in dry, grassy areas likely to become water-filled in spring. Male may contact-guard, but usually hovers nearby, especially in cool weather. Overwinters in egg diapause, hatching in spring with rising temperatures and water levels.

Striped Meadowhawk *Sympetrum pallipes*

APRIL	MAY	JUNE	JULY	AUG	SEPT	OCT

Breeding habitat: Permanent or temporary ponds, shallow marshes, fens and ditches, often sedge-rimmed.

Nature Notes:

One of many *Sympetrum* species often found at high elevations, especially in southern part of its range.

Ovipositing pairs are evidently attracted to each other, as sizeable groups are often seen ovipositing in small areas.

Scientific names *pallipes* means "pale-footed," and some populations do have pale lower legs, but it is not a constant character across the species range.

Description: Adults average 1.4 inches long.

Male's thorax brown with pair of whitish, oblique side stripes (↑) and two narrower stripes on top. Abdomen red with black, forward-pointing triangles on segments 4 to 9, reduced or wanting in some populations. Face tan; eyes reddish brown; legs variable, yellow to black. Wings clear with leading edge vein pale and often with touch of amber at bases. Female patterned like male, usually brown where he is red, but sometimes becoming red. Her wings may have orange wash at bases, extending along leading edges to nodus. Her subgenital plate about half as long as segment 9 and with two tips pointing rearward.

Identification Clues: Male is only meadowhawk with obvious thoracic stripes when mature, but several other species have similar stripes when immature and on females. Of

This female shows shapes of dark markings along sides of her abdomen and pale leading edges of wing veins.

those, Red-veined Meadowhawk is most similar but has reddish wash throughout wings, lacking on Striped. Also, Striped has modified mitten-type male hamule (pg. 257). White-faced Meadowhawk somewhat similar but smaller, with white face and lacking white thoracic stripes.

Distribution: All Rocky Mountain regions in Foothills, Montane and Subalpine zones.

Life Cycle & Behavior: Adults have a long flight period, emerging in June and July and flying into October. Females oviposit in tandem or alone in shallow water or on bank-side grassy areas including lawns, sometimes dropping eggs from air. Eggs may be laid in dry, grassy areas likely to become water-filled in spring. Overwinters in egg diapause, hatching in spring with rising temperatures and water levels.

Cherry-faced Meadowhawk *Sympetrum internum*

APRIL	MAY	JUNE	JULY	AUG	SEPT	OCT

Breeding habitat: Shallow, temporary ponds, lake edges and wet meadows that dry late in summer during some years.

Nature Notes:

Another high-elevation Sympetrum, often found above 9,000 feet in Colorado.

Egg-laying occurs no earlier than mid-day, evidently allowing pairs to distinguish persistently moist ground from dew-wet grass. Sadly, they can be fooled by watered lawns.

Has disappeared from parts of its former range in West because of persistent drought.

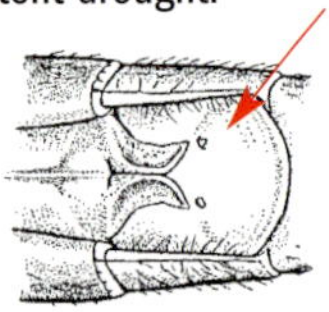

Female subgenital plate ventral view.

Description: Adults average 1.3 inches long.

Male's thorax hairy, unmarked brown. Abdomen bright red with black, forward-pointing triangles on segments 4 to 9. Face reddish (↑); eyes reddish brown; legs black; and terminal appendages yellowish. Wings with very light yellow wash throughout, pale veins along leading edges and a splash of amber at bases, variable in size. Female has markings as on male but is light brown instead of red, sometimes becoming red with age. Her eyes and face brownish and her wing stigmas may be paler at ends than in middle. Her subgenital plate about half length of segment 9, with two tips curving away from center (↑). Tenerals like female but yellow-orange.

Identification Clues: One of just two Rocky Mountain meadowhawks with an unmarked thorax, black triangles on sides of abdomen

Note light yellowish wash throughout wings of this female.

and black legs. Distinguish it from the very similar White-faced meadowhawk by the reddish face and pale costal wing veins of Cherry-faced. Also, Cherry-faced has deeply forked hamules in side view ("forked-type" pg. 257), easily separating it from all other Rocky Mountain species. Female cannot always be separated from female White-faced, but Cherry-faced has pale leading edge of wings and slight difference in shape of subgenital plates. Tips of subgenital plate of White-faced do not curve away from center.

Distribution: All Rocky Mountain regions in Foothills, Montane and Subalpine zones.

Life Cycle & Behavior: Adults have an early and long flight period, emerging in June and early July and flying into September. Females oviposit in tandem or alone, sometimes in shallow water, often in grassy areas likely to become water-filled in spring. Pairs often oviposit in groups, with females dropping eggs on ground from several inches to about a foot in the air. Overwinters in egg diapause, hatching in spring when water fills habitat.

Saffron-winged Meadowhawk *Sympetrum costiferum*

APRIL	MAY	JUNE	JULY	AUG	SEPT	OCT

Breeding habitat: Reedy marshes and pond margins, often sandy in open landscapes; tolerates saline and alkaline waters.

Nature Notes:

Spotty distribution in Southern Rockies, but was found by Scotty Willey at over 11,000 feet in Gunnison County, Colorado.

Pioneering American odonatist Clarence Kennedy once reported thousands of adults perched on a telephone wire, all facing the same way.

One of the latest dragonflies of autumn.

Description: Adults average 1.4 inches long.

A small, late-season, all-red meadowhawk with black and pale legs. Male's thorax reddish brown and lightly marked with thin black lines. Abdomen red with thin black lines low on sides of most segments (↑) and tiny black marks on top of segments 8 and 9. Eyes and face reddish brown; wings mostly clear with reddish or orange leading edges (↑), becoming fainter in older individuals; legs bicolored, black with femur (thigh) mostly pale and tibia (shin) often striped. Stigmas red, orange or yellow with black borders front and back. Female like male, but paler: yellowish orange or tan, but sometimes reddish. Her subgenital plate very short and inconspicuous. Tenerals light yellowish orange.

Identification Clues: No other meadowhawk has exact combination of nearly

Mature female shows dark coloration.

Teneral female is paler than mature female.

unmarked thorax, abdomen mostly red with only thin black lines along sides and leading edges of wings reddish or orange. Most like Autumn Meadowhawk, but legs having some black, wings with pale leading edges and female's reduced subgenital plate easily separate it from that species. Male's hamule V-shaped (pg. 257).

Distribution: All Rocky Mountain regions in Foothills and Montane Zone.

Life Cycle & Behavior: Adults have long, late flight period, emerging in July and flying into October. Males territorial, flying and perching low and hovering often. Females oviposit in tandem, occasionally alone, usually dipping her abdomen repeatedly in shallow water, sometimes on moist shores. Oviposition usually described as deliberate and methodical. Overwinters in egg diapause, hatching in spring with rising temperatures and water levels.

Band-winged Meadowhawk

Sympetrum semicinctum fasciatum

APRIL	MAY	JUNE	JULY	AUG	SEPT	OCT

Breeding habitat: Small ponds, shallow marshes, sloughs, spring seeps and fens, often with very slight current.

Nature Notes:

Nymphs don't compete well with other dragonflies and are very vulnerable to fish predation; both factors limit its habitat options.

Often perches in open on fences and telephone wires.

Western populations of this transcontinental species complex were long considered a separate species (*Sympetrum occidentale*), but are now usually divided into three western subspecies of *S. semicinctum*, only one of which occurs in the Rockies.

Description: Adults average 1.3 inches long.

A remarkable little species with distinctive orange-brown patches on the basal halves of the wings, more pronounced on the hind wings (↑). Although these patches vary in size and intensity, the color is usually darkened below the nodus more so than at the wing base. Thorax of male is reddish brown with pattern of black lines on the sides. Abdomen is red with small black marks on sides and near tip. Eyes and face reddish brown; legs black. Female is light brown with conspicuous black lines on sides of thorax and sides of abdomen, thickening near tip. Her wing patches paler than males, but still obvious. Her subgenital plate very short.

Identification Clues: Wing banding is unique. Other meadowhawks at some stages of

Wing patches on this female are much darker below the nodus than at the wing bases.

development may have yellow, orange or brownish washes of various sizes at wing bases, but margins are typically diffuse, not with crossbands of deepened color below the nodus that are perpendicular to the wing's long axis with well-defined margins. In rare confusing cases, pattern of black lines on sides of thorax more extensive than on other red meadowhawks. Male's hamule V-shaped (pg. 257).

Distribution: All Rocky Mountain regions in Foothills and Montane Zone.

Life Cycle & Behavior: Adults emerge in July and fly into October. May range far from water to forage in fields and clearings and tends to fly and perch higher than other meadowhawks. Females oviposit in tandem, dipping down to tap abdomen on water surface, usually near emergent plants. More research is needed on its life history and phylogenetic relationships.

Black Meadowhawk *Sympetrum danae*

APRIL	MAY	JUNE	JULY	AUG	SEPT	OCT

Breeding habitat: Sedge fens, sphagnum bogs, marshes and shallow, sedge-rimmed lakeshores.

Nature Notes:

Has been found at very high altitudes in Rockies, over 11,000 feet in Colorado.

Overwintering eggs have been recorded to survive for nine weeks in a block of ice.

Weakly migratory in North America, but has been encountered in large numbers over the ocean near the Irish Coast.

The only circumboreal meadowhawk. It is common in Northern Europe, Asia and Japan.

Description: Adults average 1.2 inches long.

A small, black and yellow meadowhawk with a black face. Mature male is almost entirely black (↑). Face and legs black (↑). Females and tenerals have brown on top of thorax and complex pattern of black and yellow on sides (↑). Their abdomen is light brown above with black lateral stripes along almost entire length and small black marks on top of segments 8 and 9. Male's wings clear, female's with yellow costal margin (leading edge). Teneral males gain overall black coloration slowly and gradually, usually retaining some small yellow markings.

Identification Clues: Unmistakable—only meadowhawk without red markings. Because of his dark color and small size, mature male can be confused only with a whiteface, but all whitefaces have bright white faces. Females and tenerals have more black on sides of tho-

Hint of the complex pattern of black and yellow is visible on the side of this female's thorax.

rax and abdomen than other meadowhawks and complex black/yellow pattern on thorax side is unique. Male's hamule V-shaped (pg. 257).

Distribution: All Rocky Mountain regions in Foothills, Montane and Subalpine zones.

Life Cycle & Behavior: Adults emerge from mid-July through August and fly well into autumn. Males are not territorial, perching low on shrubs or on ground. Female oviposits in tandem or alone if few other males present. She darts quickly from place to place, dispersing eggs in small areas of open water, on moss or on mud, sometimes dropping them from the air. Overwinters in egg diapause, hatching in spring. Nymphs are presently indistinguishable from nymphs of Cherry-faced Meadowhawk.

Autumn Meadowhawk *Sympetrum vicinum*

APRIL	MAY	JUNE	JULY	AUG	SEPT	OCT

Breeding habitat: Lakes, permanent ponds and slow streams, usually well-vegetated, in forested landscapes.

Nature Notes:

Engages in a type of after-tandem courtship involving false egg-laying movements prior to mating and real egg-laying.

Formerly called the Yellow-legged Meadowhawk. Though focusing on the pale legs is key to identifying it, the name was misleading because legs are often tan or reddish.

Very cold-tolerant, feeding at temperatures in upper 40s F and surviving light frosts. Flies later into autumn than any other dragonfly in northern part of its range.

Description: Adults average 1.3 inches long.

A small, late-season, all-red meadowhawk with pale legs (↑). Male nearly all red with small black spot on top of abdominal segment 9 (↑). Lacks black triangles on sides of abdomen characteristic of related species, though may have small black streaks there. Face red; wings clear; legs tan or reddish brown, lacking any black. Female like male, but paler: reddish, yellowish orange or tan. Her legs light brown to yellow. Her subgenital plate prominent and spout-shaped (↑).

Identification Clues: Easy to identify. Male is unique with pale legs lacking any black, almost entirely red body color with a small black spot on segment 9 and clear wings (sometimes tiny amber areas at hind wing bases). Female paler but distinctive for same reasons and her prominent, spout-shaped

Female's spout-shaped subgenital plate is clearly visible in side view.

subgenital plate is diagnostic in side view. Male's hamule V-shaped (pg. 257).

Distribution: Abundant in most of its transcontinental range, records in the Rockies are few and scattered in Foothills and Montane Zone of all regions.

Life Cycle & Behavior: Adults emerge very late in season, in late July and August, and fly deep into autumn. Males are not territorial. Pairs oviposit while hovering in tandem, often near other ovipositing pairs. Female repeatedly dips her abdomen into shallow water near shore or onto wet vegetation or mud, sometimes alternating between. Overwinters as egg, hatching in spring with arrival of favorable temperatures and rising water levels.

Wandering Glider *Pantala flavescens*

APRIL	MAY	JUNE	JULY	AUG	SEPT	OCT

Breeding habitat: Small, shallow standing waters, including temporary and artificial pools and ditches, usually in open areas.

Nature Notes:

The strongest migrant among the Odonata, well known around the world for crossing thousands of miles of ocean expanses in its travels, flying at night and during day.

The world's only truly cosmopolitan dragonfly, breeding on all continents except Europe and Antarctica. It is the only dragonfly on many tropical oceanic islands.

Feeds at about head height, sometimes in swarms.

Description: Adults average 1.9 inches long.

Thorax is unmarked brownish or grayish yellow. Abdomen yellow on sides, becoming orange on top, with dark median line widening to spots toward tip. Eyes reddish, face orange, wings clear with hind wings having huge surface areas (↑). Female very similar to male in color and shape, but slightly paler overall.

Identification Clues: Large size, clear wings with extraordinarily large hind wing surface area and top-striped abdominal pattern separate this species from most other dragonflies. Most like Spot-winged Glider, but is lighter in color and lacks the small dark spots at rear wing bases possessed by Spot-winged. Looks yellow to yellowish orange in its powerful, gliding flight, with tapered abdomen often apparent.

Females look very similar to males in both color and shape.

Distribution: Uncommon in all Rocky Mountain regions mostly on western slope of the Continental Divide in Foothills and lower Montane Zone.

Life Cycle & Behavior: Territorial and feeding flights are rapid and sustained, males coursing back and forth over or near water. Populations in the Rockies are likely sustained by adults migrating from south in the spring. They mate shortly after arrival here and may fly into September. Pairs oviposit in tandem, females tapping water surface every few feet. Nymphs complete development in as little as a month; migrate south shortly after they emerge in autumn. Eggs laid in inappropriate waters or too late in season may result in part-grown nymphs unable to survive winter.

Spot-winged Glider *Pantala hymenaea*

APRIL	MAY	JUNE	JULY	AUG	SEPT	OCT

Breeding habitat: Small, shallow standing waters, including temporary and artificial wetlands and pools, usually in open areas.

Nature Notes:

My best collection of nymphs of this species came from a small, bright blue cement pool at a miniature golf course. Shutting down his operation in late fall, the owner noticed a bunch of "bugs" crawling on the bare bottom and contacted a naturalist friend of mine.

Strongly migratory with powerful flight, but doesn't seem to cross ocean expanses like the Wandering Glider. Has been found in desert 50 miles from nearest water.

This powerful, agile, erratic flier can be very challenging to net or photograph.

Description: Adults average 1.9 inches long.

Thorax is brown to reddish brown with two pale gray stripes on sides. Abdomen light brown on sides, darkening on top, with median black line widened to spots on some segments. Eyes red above, gray below; face red; wings clear with huge hind wing surface areas (↑) and dark oval spots at bases (↑). Female very similar to male in color and shape, her face often brown or orange.

Identification Clues: Large size, mostly clear wings with large hind wing surface areas and complex abdominal pattern with black line and spots on top, separate this species from most other dragonflies. Most like Wandering Glider, but is darker and possesses dark, oval spots at hind wing bases that can usually be seen in flight.

The oval hind wing spots, like the one visible on this female, are prominent enough to be seen in flight.

Distribution: Uncommon in all Rocky Mountain regions mostly on western slope of the Continental Divide in Foothills and lower Montane Zone.

Life Cycle & Behavior: Feeding and territorial flights are very swift and erratic. Males chase large dragonflies of their own and other species from prime areas. Often feeds in swarms. Populations in the Rockies are likely sustained by adults migrating from south in the spring. They mate shortly after arrival and fly into autumn. Female often oviposits alone, sometimes in tandem; if in tandem pair flies erratically, plunging suddenly to deposit eggs in a single dip, then moving elsewhere to repeat. Nymphs complete development in five weeks or less; migrate south shortly after they emerge in autumn.

Red Saddlebags *Tramea onusta*

APRIL	MAY	JUNE	JULY	AUG	SEPT	OCT

Breeding habitat: Primarily ponds, including temporary ones, also lakes, ditches, canals and slow rivers.

Nature Notes:

Likely does best in fish-free ponds and pools, much as rainpool gliders do.

In addition to engaging in TTG, females sometimes oviposit alone, then behaving differently with slow, low flight and much tapping into algal mats and other moist vegetation.

The pattern of red veins in the dark wing patches is striking.

Description: Adults average 1.8 inches long.

Male's thorax mostly unmarked reddish brown. Abdomen red with black spots on top of segments 8 and 9. Eyes and face red. Wings red-veined at bases and along costal margins; fore wings clear; hind wings each with basal area holding a large, reddish brown patch with irregular margins (↑). Females and tenerals similar to males, but brownish.

Identification Clues: Unmistakable. Because of the shape of the hind wing patch and lack of any pruinosity, it is unlikely to be confused with another dragonfly genus. Most like Black Saddlebags but is red, brownish if female, but not black. Quite similar to the Carolina Saddlebags, which is not known to occur in the Rockies, but as with any wide-ranging species, unusual looking individuals should be examined carefully.

Red saddlebag-shaped hind wing patch is the same on both males and females.

Distribution: Uncommon in scattered locations in Foothills and lower Montane Zone of Southern Rockies and southern tip of Central Rockies.

Life Cycle & Behavior: Often feeds in mixed swarms with Black Saddlebags and rainpool gliders, sometimes far from water. Males patrol large territories of about 300 square feet with much energy, always ready to confront intruders. Populations in the Rockies are likely sustained by adults migrating from south in the spring, although this is not certain. Pairs engage in TTG when ovipositing (see genus introduction), but females also oviposit alone (see Nature Notes). Adults fly into autumn, then likely head south. Little is known about the life history of the nymphs, except that they can grow fast.

Black Saddlebags *Tramea lacerata*

APRIL	MAY	JUNE	JULY	AUG	SEPT	OCT

Breeding habitat: Primarily shallow, heavily vegetated ponds, including temporary ones, also some open lakeshores.

Nature Notes:

Likely does best in fish-free or low-fish environments.

Dennis Paulson observed that black bass will follow ovipositing pairs underwater, ready to strike when female descends to water to oviposit.

Scientific name *lacerata* means "lacerated," likely alluding to the ragged edges of the hind wing patches.

Description: Adults average 2.1 inches long.

Male's thorax unmarked and blackish. Abdomen black with a large, yellow or white, squarish spot on segment 7 that fades with maturity. Spot usually remains faintly visible. Eyes and face dark. Fore wings clear; hind wings each with basal quarter holding a large, black patch with irregular margins (↑). Females similar to males but duller, blackish to dark gray. Tenerals are dark brown, with large, yellow, squarish spot on segment 7 and some preceding segments.

Identification Clues: Unmistakable. Because of the shape of the hind wing patch and lack of any pruinosity, it is unlikely to be confused with another dragonfly genus. Most like Red Saddlebags but is black, not red, including female. The often prominent, pale, squarish spot on segment 7—and on preceding segments of immatures—also distinctive.

Saddlebags engage in a unique form of guarded egg-laying (called TTG—see also pg. 221). Tandem flight (top) is interrupted as the male releases the female (bottom); she drops down to water to tap her abdomen tip, releasing eggs, then quickly ascends to re-engage with the guarding male. Pair continually repeats the process.

Distribution: Uncommon in Foothills and lower Montane Zone along western edges of Northern and Central Rockies and eastern edge of Southern Rockies.

Life Cycle & Behavior: Feeding and territorial flights are bouncy and somewhat erratic, effortless gliding interspersed with speedy bursts. Males defend shoreline territories of about 100 feet. Often feeds in mixed-species swarms, sometimes far from water over meadows and hilltops. Populations in the Rockies are sustained by adults migrating from south in the spring. They mate shortly after arrival and persist into autumn. Pairs engage in TTG when ovipositing (see genus introduction and photos above). Nymphs develop quickly—though not as quickly as rainpool gliders in the lab—and migrate south shortly after they emerge in autumn.

Glossary

Anal loop: Group of cells in the hind wing of Anisoptera that is boot-shaped or vaguely foot-shaped in skimmers and emeralds.

Anisoptera: Suborder of Odonata containing the dragonflies.

Antehumeral stripe: Pale stripe on side of the thorax between the dark mid-dorsal stripe and the dark humeral stripe of some damselflies. Commonly called the pale shoulder stripe.

Apical: Directional term meaning at or near the tip of a structure.

Basal: Directional term meaning at or near the base of a structure.

Cerci: (singular: cercus) Upper pair of the terminal appendages (claspers) of the male. Only prominent terminal appendage of some females (a.k.a paddles).

Claspers: Common term for the terminal appendages (made up of superior and inferior appendages) of the male that are used to grasp the thorax or rear of the head of the female during mating (a.k.a. caudal appendages). Dragonfly claspers consist of the cerci and the epiproct. Damselfly claspers consist of the cerci and the paraprocts.

Dorsal: Directional term meaning on the top or above.

Emergence: When the mature nymph leaves the water and transforms (undergoes metamorphosis) into an adult. Leaves the cast exoskeleton (exuvia) behind.

Endophytic oviposition: Laying eggs inside of plant tissue.

Epiproct: Lower structure of terminal appendages (claspers) of the male Anisoptera (dragonflies).

Exuviae: Shed skin left behind when the nymph climbs out of the water and transforms into the adult (the Latin source word has only the plural form, but many odonatists use a singular form exuvia to avoid ambiguity).

Frons: Upper part of odonate head, popularly called the "forehead."

Hamules: (a.k.a hamular processes) Secondary genitalia of the male comprising a pair of structures on the underside of abdominal segment 2. Holds female genitalia in place during copulation.

Heterochromatic: "Typical" female coloration not resembling the male (a.k.a heteromorph).

Homeochromatic: Female coloration resembling male (a.k.a andromorph).

Humeral stripe: Dark stripe on side of thorax below the pale antehumeral stripe of some damselflies. Commonly called the dark shoulder stripe.

In copula: ("in cop" for short) A pair of odonates engaged in the wheel-shaped mating configuration (also see Wheel).

Instar: A nymphal stage between molts, most species progress through about 12 before maturation.

Labium: The outer, lower part of the mouth, sometimes thought of as the "chin" and lower "lip."

Labrum: The outer, upper part of the mouth, sometimes thought of as the upper "lip."

Lateral: Directional term meaning on the side(s).

Mesostigmal plates: Small sclerites at the front and top of the pterothorax of damselflies that are often modified in females for contact with the male claspers when in tandem. Sometimes called "shoulder pads."

Middorsal stripes: Stripes on damselflies that are usually dark, flanking the middorsal carina (longitudinal ridge running down the middle of the thorax) on each side.

Nodus: Shallow notch on the leading edge of the wing where the two anterior wings veins (the costa and the subcosta) come together.

Nymph: Immature aquatic stage of the odonate. Sometimes called larva or naiad.

Occipital bar: Pale, transverse line found along the back of the head of some damselflies between the postocular spots (eyespots).

Obelisking: Pointing the tip of the abdomen toward the sun to cool by reducing the surface area exposed to sunlight.

Ocelli: Three simple eyes found on top of an adult odonate's head, (on vertex).

Occiput: Top of head behind eyes.

Odonata: Biological order of insects containing the dragonflies and damselflies. The term odonate is often used to describe members of this order.

Odonatist/Odonatologist: A person who studies dragonflies and damselflies.

Oviposition: The act of laying eggs, which the female may do alone or accompanied by the male.

Ovipositor: The structure on the bottom side of the abdomen at segments 8 and 9 of female damselflies and darners that they use to pierce an appropriate piece of living or dead plant material and deposit eggs.

Paraprocts: The lower pair of terminal appendages (claspers) of the male damselfly (a.k.a inferior appendages).

Pharate adult: Newly formed adult that has not yet emerged from the nymphal skin.

Postclypeus: Middle part of the face (with small anteclypeus immediately below it).

Postocular spots: (a.k.a. eyespots) Pale colored spots on top of the head near the eyes of damselflies, sometimes connected to each other by a thin bar (occipital bar).

Prothorax: The first segment of the thorax, which has no wings but has the first pair of legs.

Pruinosity: A whitish, grayish or pale bluish, waxy substance (bloom) that exudes from the cuticle as some odonates mature. It may cover much of the body of an older adult (a.k.a. pruinescence).

Pseudostigma: Area of contrasting color at front tip of the wing near where a true stigma would be found on the Calopterygidae, but often larger than a true stigma and containing multiple cells.

Pterothorax: The combined middle (meso-) and hind (meta-) parts of the thorax that are fused together for strength. Bears the wings and two pairs of legs (a.k.a. synthorax).

Stigma: Single, blood-filled cell near the tip of the wing that contrasts in color and texture with the rest of the wing (a.k.a. pterostigma).

Subgenital plate: Plate of variable shape on underside of abdominal segment 8 of females of many odonate species that holds the eggs prior to release and protects the genital opening.

Suture: Line of contact, often furrowed, where two plates of the exoskeleton meet on the face or thorax.

Synthorax: = pterothorax.

Tandem or Tandem Linkage: Position in which the male is holding the thorax or rear of head of the female with his terminal appendages (claspers), but the pair is not copulating.

Tubercle: a small, knob-like "bump."

Teneral: young adult odonate just after it has emerged from the nymphal stage but before its wings and sclerites are fully hardened and it is full colored (not yet sexually mature).

Ventral: directional term meaning below or underneath.

Vertex: Cone-like protuberance on center of head in front of eyes on which the ocelli are located.

Vulvar lamina: = subgenital plate.

Vulvar spine: Sharp, pointed extension at the bottom end of abdominal segment 8 on females of some pond damsels.

Wheel: The circular or heart-shaped position of mating odonates following tandem linkage with the female engaging the male's secondary genitalia with the tip of her abdomen.

Zygoptera: The suborder of Odonata containing the damselflies.

Appendix A

Checklist of Damselflies & Dragonflies of the Rocky Mountains

		Glacier NP	Rocky Mtn. NP	Yellowstone/ Grand Teton
Broad-winged Damselflies (Family Calopterygidae)				
❑ *Calopteryx aequabilis*	River Jewelwing	❑		❑
❑ *Hetaerina americana*	American Rubyspot			
Spreadwings (Family Lestidae)				
❑ *Archilestes grandis*	Great Spreadwing		❑	
❑ *Lestes congener*	Spotted Spreadwing	❑	❑	❑
❑ *Lestes disjunctus*	Northern Spreadwing	❑	❑	❑
❑ *Lestes forcipatus*	Sweetflag Spreadwing	❑		
❑ *Lestes unguiculatus*	Lyre-tipped Spreadwing	❑	❑	❑
❑ *Lestes dryas*	Emerald Spreadwing	❑	❑	❑
Pond Damsels (Family Coenagrionidae)				
❑ *Argia apicalis*	Blue-fronted Dancer			
❑ *Argia moesta*	Powdered Dancer			
❑ *Argia sedula*	Blue-ringed Dancer		❑	
❑ *Argia alberta*	Paiute Dancer			❑
❑ *Argia emma*	Emma's Dancer	❑	❑	❑
❑ *Argia fumipennis violacea*	Violet Dancer	❑	❑	
❑ *Argia plana*	Springwater Dancer			
❑ *Argia vivida*	Vivid Dancer		❑	❑
❑ *Argia lugens*	Sooty Dancer			
❑ *Coenagrion interrogatum*	Subarctic Bluet	❑		
❑ *Coenagrion resolutum*	Taiga Bluet	❑	❑	❑
❑ *Enallagma praevarum*	Arroyo Bluet		❑	❑
❑ *Enallagma anna*	River Bluet	❑	❑	❑
❑ *Enallagma civile*	Familiar Bluet	❑	❑	
❑ *Enallagma carunculatum*	Tule Bluet	❑	❑	❑
❑ *Enallagma clausum*	Alkali Bluet	❑	❑	❑
❑ *Enallagma annexum*	Northern Bluet	❑	❑	❑
❑ *Enallagma boreale*	Boreal Bluet	❑	❑	❑
❑ *Enallagma ebrium*	Marsh Bluet	❑		❑
❑ *Enallagma hageni*	Hagen's Bluet			
❑ *Enallagma basidens*	Double-striped Bluet		❑	

Checklist of Rocky Mountain Odes

Species	Common Name	Glacier NP	Rocky Mtn NP	Yellowstone/ Grand Teton
❑ *Ischnura barberi*	Desert Forktail			
❑ *Ischnura cervula*	Pacific Forktail	❑	❑	❑
❑ *Ischnura damula*	Plains Forktail		❑	
❑ *Ischnura demorsa*	Mexican Forktail		❑	
❑ *Ischnura perparva*	Western Forktail	❑	❑	❑
❑ *Ischnura verticalis*	Eastern Forktail		❑	
❑ *Ischnura denticollis*	Black-fronted Forktail			
❑ *Ischnura hastata*	Citrine Forktail			
❑ *Hesperagrion heterodoxum*	Painted Damsel		❑	
❑ *Amphiagrion abbreviatum*	Western Red Damsel	❑	❑	❑
❑ *Nehalennia irene*	Sedge Sprite	❑		

Darners (Family Aeshnidae)

Species	Common Name	Glacier NP	Rocky Mtn NP	Yellowstone/ Grand Teton
❑ *Aeshna tuberculifera*	Black-tipped Darner	❑		
❑ *Aeshna juncea*	Sedge Darner	❑	❑	❑
❑ *Aeshna subarctica*	Subarctic Darner	❑		
❑ *Aeshna canadensis*	Canada Darner	❑		
❑ *Aeshna eremita*	Lake Darner	❑	❑	❑
❑ *Aeshna interrupta interna*	Variable Darner	❑	❑	❑
❑ *Aeshna sitchensis*	Zigzag Darner	❑		❑
❑ *Aeshna constricta*	Lance-tipped Darner	❑	❑	❑
❑ *Aeshna palmata*	Paddle-tailed Darner	❑	❑	❑
❑ *Aeshna umbrosa occidentalis*	Shadow Darner	❑	❑	❑
❑ *Rhionaeschna californica*	California Darner	❑	❑	
❑ *Rhionaeschna multicolor*	Blue-eyed Darner	❑	❑	❑
❑ *Anax junius*	Common Green Darner	❑	❑	❑

Clubtails (Family Gomphidae)

Species	Common Name	Glacier NP	Rocky Mtn NP	Yellowstone/ Grand Teton
❑ *Gomphus externus*	Plains Clubtail			
❑ *Gomphus graslinellus*	Pronghorn Clubtail	❑		
❑ *Stylurus intricatus*	Brimstone Clubtail			
❑ *Ophiogomphus occidentis*	Sinuous Snaketail	❑		
❑ *Ophiogomphus severus*	Pale Snaketail	❑	❑	❑

Spiketails (Family Cordulegastridae)

Species	Common Name	Glacier NP	Rocky Mtn NP	Yellowstone/ Grand Teton
❑ *Cordulegaster dorsalis*	Pacific Spiketail	❑		

Cruisers (Family Macromiidae)

Species	Common Name	Glacier NP	Rocky Mtn NP	Yellowstone/ Grand Teton
❑ *Macromia magnifica*	Western River Cruiser			

Emeralds (Family Corduliidae)

Species	Common Name	Glacier NP	Rocky Mtn NP	Yellowstone/ Grand Teton
❑ *Cordulia shurtleffii*	American Emerald	❑	❑	❑

Checklist of Rocky Mountain Odes

		Glacier NP	Rocky Mtn NP	Yellowstone/ Grand Teton
❑ *Somatochlora albicincta*	Ringed Emerald			
❑ *Somatochlora hudsonica*	Hudsonian Emerald	❑	❑	❑
❑ *Somatochlora semicircularis*	Mountain Emerald	❑	❑	❑
❑ *Somatochlora minor*	Ocellated Emerald	❑	❑	❑
❑ *Somatochlora walshii*	Brush-tipped Emerald	❑		
❑ *Somatochlora ensigera*	Plains Emerald			
❑ *Epitheca spinigera*	Spiny Baskettail	❑		

Skimmers (Family Libellulidae)

		Glacier NP	Rocky Mtn NP	Yellowstone/ Grand Teton
❑ *Plathemis subornata*	Desert Whitetail			
❑ *Plathemis lydia*	Common Whitetail	❑	❑	
❑ *Ladona julia*	Chalk-fronted Corporal	❑		
❑ *Libellula quadrimaculata*	Four-spotted Skimmer	❑	❑	❑
❑ *Libellula saturata*	Flame Skimmer			❑
❑ *Libellula composita*	Bleached Skimmer			
❑ *Libellula nodisticta*	Hoary Skimmer			
❑ *Libellula forensis*	Eight-spotted Skimmer		❑	❑
❑ *Libellula pulchella*	Twelve-spotted Skimmer	❑	❑	❑
❑ *Libellula luctuosa*	Widow Skimmer		❑	
❑ *Leucorrhinia proxima*	Belted Whiteface	❑	❑	❑
❑ *Leucorrhinia intacta*	Dot-tailed Whiteface	❑	❑	❑
❑ *Leucorrhinia hudsonica*	Hudsonian Whiteface	❑	❑	❑
❑ *Leucorrhinia glacialis*	Crimson-ringed Whiteface	❑		❑
❑ *Leucorrhinia borealis*	Boreal Whiteface		❑	❑
❑ *Erythemis collocata*	Western Pondhawk		❑	❑
❑ *Sympetrum corruptum*	Variegated Meadowhawk	❑	❑	❑
❑ *Sympetrum illotum*	Cardinal Meadowhawk			
❑ *Sympetrum madidum*	Red-veined Meadowhawk	❑		❑
❑ *Sympetrum obtrusum*	White-faced Meadowhawk	❑	❑	❑
❑ *Sympetrum pallipes*	Striped Meadowhawk	❑	❑	❑
❑ *Sympetrum internum*	Cherry-faced Meadowhawk	❑	❑	❑
❑ *Sympetrum costiferum*	Saffron-winged Meadowhawk	❑	❑	❑
❑ *Sympetrum semicinctum*	Band-winged Meadowhawk	❑	❑	❑
❑ *Sympetrum danae*	Black Meadowhawk	❑	❑	❑
❑ *Sympetrum vicinum*	Autumn Meadowhawk		❑	
❑ *Pachydiplax longipennis*	Blue Dasher			❑
❑ *Pantala flavescens*	Wandering Glider		❑	
❑ *Pantala hymenaea*	Spot-winged Glider			
❑ *Tramea onusta*	Red Saddlebags		❑	
❑ *Tramea lacerata*	Black Saddlebags			

Appendix B & C

Odonata Groups & Websites

Many states have websites with information about their odonates. Simply google the state's name and "odonata" or "dragonflies."

Dragonflies of the Southwest **www.southwestdragonflies.net**

New Mexico Dragonfly Notebook **google "new mexico dragonflies"**

Wild Utah Dragonflies & Damselflies **www.wildutah.us**

Worldwide Dragonfly Association **google "world dragonfly"**
Publishes the *International Journal of Odontology* and a newsletter, *Agrion.*

Dragonfly Society of the Americas **www.odonatacentral.org**
DSA publishes the journal *American Odonatology* and a newsletter, *Argia.*

International Odonata Research Institute **www.iodonata.net**
Source for collecting resources and dragonfly books.

Photo Credits

T=top R=right
M=middle L= left
B=bottom i=inset

Karl Bardon: 212, 273

Giff Beaton [giffbeaton.com]: 282 T

Ryan Brady [pbase.com/rbrady]: 143, 146 C, 147 E, 148, 149 B, 152 ALL, 153, 154 TR/B, 155, 158, 163 B, 166 T, 167, 198, 199 T, 206, 207, 209, 245, 272

Robert DuBois: 28 M/B, 30, 33

Sid Dunkle: 97 T, 118, 182, 191, 193, 200, 202, 210

Jim Johnson: 285 T/B

Kurt Mead: 4 B, 26 T, 146 E, 147 A/B, 156 ALL, 160 ALL, 227 B, 249, 284

Dennis Paulson: 4 T, 63, 83, 96, 119, 120, 121, 127, 133, 135, 146 B, 147 D/F/G, 150, 151, 159, 164 T, 165 ALL, 168, 169, 170 T , 171 TR, 179, 180, 181, 183, 186, 187, 192, 204, 205, 208, 223, 231, 232, 233, 234, 235, 236, 237, 247 T, 250, 251, 261, 263, 275, 278, 279, 280, 281, 283

Mike Reese [wisconsinbutterflies.org]: 61, 67, 79, 89 B, 102, 103, 105 T, 107, 117 ALL

Sparky Stensaas [sparkyphotos.com]: COVER BM, 3, 4 M, 7, 12, 13, 15, 16, 20 BL/BR, 21 BL/BR, 24 ALL, 26 B, 27, 29 ALL, 41, 48 ALL, 49, 58 B, 60 B, 81 i, 98 B, 104, 139 B, 146 A/D, 147 C, 149 T, 161, 162 ALL, 163 T, 164 B, 166 B, 172, 173 ALL, 224, 225 ALL, 226 ALL, 227 TR, 228, 229 ALL, 238, 239 T/B, 241 ALL, 242, 243 ALL, 244, 246, 247 BL/BM/BR, 254 B, 255 T, 258 T, 259, 265, 271 B, 277, 282 B

Ken Tennessen: COVER TOP/BL, 23, 50, 51, 56, 57, 58 T, 59, 60 T, 62, 64, 65, 66, 78, 80, 81 T, 82, 84, 85, 86. 87, 88, 89 T, 90 ALL, 91, 92, 93, 94, 95, 98, 99, 100, 101, 105 B, 106, 108, 109, 110, 111, 112, 113, 114, 115, 116, 122, 123, 124, 125 ALL, 126, 128, 129, 130, 131 ALL, 132, 134, 136, 137, 138, 139 T/M, 140, 141, 154 TL, 157, 170 B, 171 TL, 178, 184, 185, 190, 199 i, 201, 213, 222, 227 TL, 230, 239 T, 240, 248, 252, 253, 254 T, 255 i, 258 B, 260, 262, 264, 266, 267, 268, 269, 270, 271 T, 274, 276

Appendix D

Titles of Interest

Abbott, J. 2005. *Dragonflies and Damselflies of Texas and the South-Central United States.* Princeton: University of Princeton Press.

Acorn, J. 2004. *Damselflies of Alberta: Flying Neon Toothpicks in the Grass.* Edmonton, AB: The University of Alberta Press.

Biggs, K. 2004. *Common Dragonflies of the Southwest, A Beginner's Pocket Guide.* Sebastopol, CA: Azalea Creek Publishing.

Brooks, S. 2003. *Dragonflies.* Washington, DC: Smithsonian Books.

Cannings, R. A., and K. M. Stuart. 1977. *The Dragonflies of British Columbia.* B. C. Provincial Museum Handbook 35. 254 pp.

Corbet, P. S. 1999. *Dragonflies, Behavior and Ecology of Odonata.* Ithaca, NY: Cornell University Press.

DuBois, B. 2005. *Damselflies of the North Woods.* Duluth, MN: Kollath+Stensaas Publishing.

Dunkle, S. 2000. *Dragonflies through Binoculars.* New York: Oxford University Press.

May, M., and S. Dunkle. 2007. *Damselflies of North America, Color Supplement.* Gainesville: Scientific Publishers.

Mead, K. 2009 (2nd edition). *Dragonflies of the North Woods.* Duluth, MN: Kollath+Stensaas Publishing.

Needham, J. G., M. J. Westfall and M. L. May. 2000. *Dragonflies of North America.* Gainesville, FL: Scientific Publishers, Inc.

Nikula, B., and J. Sones. 2002. *Stokes Beginner's Guide to Dragonflies.* Boston, MA: Little, Brown and Company.

Paulson, D. R. 1999. *Dragonflies of Washington.* Seattle: Seattle Audubon Society.

Paulson, D. R. 2009. *Dragonflies and Damselflies of the West.* Princeton, NJ: Princeton University Press.

Paulson, D. R. and S. W. Dunkle. 1999. *A Checklist of North American Odonata.* Tacoma, WA: Slater Museum of Natural History, University of Puget Sound, Occasional Paper #56.

Shaw, J. 1987. *John Shaw's Closeups in Nature.* New York, NY: Amphoto.

Walker, E. M. 1953-1975. *The Odonata of Canada and Alaska* (3 volumes). Toronto, ON: Univ. Toronto Press. Reproduced in 2000 by Toronto Entomologists Association.

Westfall, M. J. and M. L. May. 2006 (2nd edition). *Damselflies of North America.* Gainesville, FL: Scientific Publishers.

Appendix E

Binoculars for "Dragonflying"

Quality close-focusing optics have opened the door to the hobby of odonate watching or "dragonflying." Close-focusing is a must; full-size is preferable; power is a matter of choice. Listed here is a sampling of the best binoculars for watching our "winged dragons" and "winged damsels."

Weight is listed because some folks prefer lighter optics. The compacts are lighter, but they gather less light at dawn and dusk. They are also harder to hold steady than their full-size cousins. Compact binoculars are nice for easy packing on longer hikes.

Good dragonflying binoculars should focus under ten feet and preferably under six feet. This gives you the magnification to fill your field-of-view with the dragonfly.

The last column represents relative brightness. The higher the number the brighter your view will be. This is important for picking out details dusk. Some dragonflies do fly late when a brighter view would be a bonus.

What is the difference between 10-power and 8-power? With increased power one loses brightness. As stated above, this may not be a big problem with dragonflying. Also with higher power binoculars, any body movement is magnified.

Brand	Size	Weight (in oz.)	Closest focus (ft.)	Brightness
Bausch & Lomb				
8x42 WP Elite	full-size	29	5	28
10x42 WP Elite	full-size	28	5	18
10x42 WP Discoverer	full-size	30	8	18
Brunton				
8.5x43 Epoch	full-size	25	3	24
10.5x43 Epoch	full-size	25	3	18
8x21 Epoch	compact	13	6	6
8x25 Eterna	compact	12	6	8
8x32 Eterna	full-size	25	4	16
10x32 Eterna	full-size	25	4	10
7x40 Macroscope	monocular	11	1.5	??
8x32 Echo	full-size	20	6	16
10x32 Echo	full-size	20	6	10
Bushnell				
8x42 Legend roof prism	full-size	30	6	28
10x42 Legend roof prism	full-size	30	6	18

Brand	Size	Weight (in oz.)	Closest focus (ft.)	Brightness
Eagle				
8x42 Ranger Platinum	full-size	23	5.2	28
10x42 Ranger Platinum	full-size	23	5.2	18
6x32 Ranger Platinum	full-size	19	3	28
8x32 Ranger Platinum	full-size	19	3	14
Kahles				
8x42 DCF	full-size	26	8	28
8x32 DCF	full-size	22	5	14
Leica				
8x32 Ultravid BR	full-size	20	7	16
10x32 Ultravid BR	full-size	21	7	10
8x20 Ultravid BR/BL	compact	9	7	6
Minox				
8x32 BD/BR	full-size	22	5	16
Nikon				
8x32 Venturer LX	full-size	25	8	16
10x32 Venturer LX	full-size	25	8	10
8x42 Monarch ATB/PC	full-size	21	8	28
10x42 Monarch ATB/PC	full-size	21	8	18
Pentax				
8x32 DCF SP	full-size	23	5	16
8x43 DCF SP	full-size	25	7	29
10x43 DCF SP	full-size	25	7	18
6.5x21 Papilio	compact	10	1.6	10
8.5x21 Papilio	compact	10	1.6	6
Swarovski				
8x20 B/G	compact	8	8	6
8.5x42 EL	full-size	29	8	24
10x42 EL	full-size	28	8	18
Swift				
7x36 Eaglet 825R	full-size	21	5	26
10x42 Viceroy 830	full-size	22	7	18
8x42 Ultra Lite 929	full-size	22	6	28
10x42 Ultra Lite 930	full-size	24	6	18
Zeiss				
10x40 Victory	full-size	26	8	16

Index

Field Notes

Field Notes

Quick Odonata Finder